Absolutely Essential

Basic Bioethics

Arthur Caplan, editor

A complete list of the books in the Basic Bioethics series appears at the back of this book.

Absolutely Essential

Bioethics and the Rules-Based International Order

Jonathan D. Moreno

The MIT Press
Cambridge, Massachusetts | London, England

The MIT Press
Massachusetts Institute of Technology
77 Massachusetts Avenue, Cambridge, MA 02139
mitpress.mit.edu

The MIT Press would like to thank the anonymous peer reviewers who provided comments on drafts of this book. The generous work of academic experts is essential for establishing the authority and quality of our publications. We acknowledge with gratitude the contributions of these otherwise uncredited readers.

This book was set in Stone Serif and Stone Sans by Westchester Publishing Services. Printed and bound in the United States of America.

Library of Congress Cataloging-in-Publication Data is available.

ISBN: 978-0-262-55337-7

10 9 8 7 6 5 4 3 2 1

EU Authorised Representative: Easy Access System Europe, Mustamäe tee 50, 10621 Tallinn, Estonia | Email: gpsr.requests@easproject.com

For Jay Schulkin

Contents

Preface

> The voluntary consent of the human subject is absolutely essential.
>
> —Nuremberg Military Tribunal, 1947

Buried in a 127-page judgment in a trial of twenty-three Nazi doctors and medical bureaucrats in 1947, ten principles for the ethics of human experiments have gone on to have a life of their own. Known to posterity as the Nuremberg Code, the first sentence is notable for the term voluntary consent, a condition for human experiments that the judges in the trial said they considered to be absolutely essential. Though often honored mainly in the breach, consent is now a cornerstone of bioethics—the study and application of moral values in health care and the medical sciences. Consent is the focus of a vast analytical and empirical literature. By contrast, the assertion that consent is absolutely essential has elicited virtually no commentary: Essential for what and for whom? Those questions are the subject of this book.

A full understanding of the significance of the Code demands attention to its historical context. After World War II the victorious allies aimed to establish a new global framework of rules that would prevent another catastrophic conflict, one that would be still more perilous, perhaps threatening the survival of human civilization. In that new framework, the sovereignty of self-governing, democratic states must be inviolable, with the absolutely essential nature of consent—here specified in the setting of medical experiments on human beings—recognized as a core moral standard that should govern relations among nations as well as persons. The physician-patient relationship, characterized by both formal structures and informal interactions, and by a dynamic of vulnerability and power, is a kind of crucible for this essential value, especially as it pertains to the expansion of knowledge intended to enhance human flourishing through medical science.

Bioethics in the Rules-Based Order

I maintain three principle claims in this book. First, I want to argue that the field known as bioethics is a creature of what political scientists call the post–World War II liberal international order, where liberal refers to Enlightenment values like individual rights and the consent of the governed, or simply the rules-based order.[1] According to the rules-based order, international relations are to be organized according to principles of open markets, liberal democracy, and multilateral organizations. This concept is not universally accepted. Criticisms of the rules-based order range from the claim that it is illusory, a position taken by critics known as realists, to the argument that it has enabled the United States to be a global hegemon. According to both criticisms the basic issue is power, not rules.[2]

The defenders of the idea of the rules-based order are prepared to grant an element of power in the creation and maintenance of the system that has largely prevailed for nearly eight decades, but they view the return to an older imperial system as a return to the instability that led to the World Wars. Advocates of the rules-based order believe that, despite appearances to the contrary, it maintains a firm hold on international relations. In the words of a prominent theorist of the postwar system, "Although America's hegemonic position may be declining, the liberal international characteristics of order—openness, rules, multilateral cooperation—are deeply rooted and likely to persist."[3] As will be seen, others are not so sure. I will not try to resolve this vigorous debate among political scientists—though for liberal internationalists World War I was also hard to imagine until 1914—but I do maintain that whether the liberal order was actual, aspirational or mythological, bioethics has nonetheless been a part of it.

Second, I want to establish that, under a conception of the rules-based order, the life sciences in particular have reached a de facto understanding with the liberal state that matters of ethics are largely within the jurisdiction of the scientific community itself, so long as certain conditions are maintained. One of these conditions is that whatever ethics arrangements are established, political authorities are reasonably assured that they will not be subject to embarrassment by the activities of science. Another is that the security of the state or of its distinct advantages over potential adversaries must not be jeopardized by science. At that point, if the different interests of self-governing science and the political system diverge, the former will be subordinated by the latter. Bioethics has become an essential part of this understanding.

Third, I want to consider what the implications of a collapse of the liberal order, or its substantial revision, would be for bioethics. This matter obviously ties into the previous point about professional jurisdiction. Within that framework bioethics has flourished, especially as it has evolved from traditional medical ethics. The latter was largely a set of directives to medical doctors about their duties to those in their care, while modern bioethics has emphasized the role of the patient or (especially) the research subject in medical decision making. In this sense bioethics is a quite recent development, largely based on events and trends in the United States in the later 1960s and early 1970s, but one also rooted in principles established in the post–World War II trials at Nuremberg. In particular, the Nazi doctors' trial was an implicit effort to bring medical ethics under the ambit of an expansive new rules-based order that was being promulgated by the United States and its allies. As we shall see, the translation of principles into rules has been at best imperfect.

Like so much of American culture since World War II, American-style bioethics, with its emphasis on respect for autonomy, has had an outsized influence on the theory and practice of bioethics in the rest of the world. Critics of the international order have no difficulty seeing this influence as one example of the hegemonic role of the United States in the decades since World War II, a rationale or a convenient excuse for the "full spectrum dominance" sought by the US military establishment, and what critics view as Americans' often disastrous sense of national exceptionalism and moral superiority. Nor has the US been reluctant to deviate from those usually advantageous rules when it sees it necessary to do so—including bending, if not breaking, rules such as those governing the system of international trade that it helped put in place; the US has for various reasons failed to sign or ratify at least ten major treaties.[4] But it is far from the only offender against the spirit of the rules-based order. In 2024, the president of the International Committee of the Red Cross observed that the Geneva Conventions to protect civilians, detainees, and wounded soldiers, adopted by nearly all nation-states since 1949, are "under strain, disregarded, [and] undermined to justify violence."[5] Nor have certain more subversive means of imposing great power interests than outright invasions been abandoned in the alleged postimperial order, among them coups d'état and election interference. As become especially evident after the election of the 47th president, the end of the old imperial order should not be confused with the end of imperial means of imposing power.[6]

Nonetheless, advocates of the rules-based order view it as vastly superior to the older imperial order. The rules-based order is seen by its advocates as having improved both respect for human rights and the human condition. Examples include advances in scientific knowledge through international cooperation while protecting (or at least striving to protect) the rights of human beings who are involved in medical interventions—whether as participants in experiments, as patients, or as members of the public—with a moral and practical stake in the intellectual and policy products of science. Bioethics norms and procedures, including informed consent and prior review of study protocols, are baked into global drug and device research and development, but less so the global health system that is now in a precarious state. (Standards for the use of nonhuman animals in medical experiments are of course part of the bioethics literature and policies, but in this book, I will focus on bioethical issues specifically involving human beings.)

Why Bioethics?

In the public sphere too, bioethics punches above its weight. The medical ethics scandals of the 1960s and early 1970s in the US were intensely covered in a media universe that was beginning to venture more decisively into investigative reporting, just as the country itself was entering a period of skepticism about the competence and the moral compass of experts. Behavioral scientists, as well as medical experimenters, found their treatment of research participants being questioned. Civil rights movements among Black Americans, women, and people institutionalized as mentally ill characterized the era. (The rights claims of gay and transgender people would be recognized later.) A generation of young philosophers, theologians, and physicians objected to the detachment of their educational experience from the great issues of the day. Some medical schools established professorships in ethics or medical humanities, and colleges created courses with titles like Present Moral Problems that focused on cases and issues of the day rather than high moral theory. Two think tanks devoted to these issues were founded, the Hastings Center in 1969 and the Kennedy Institute of Ethics at Georgetown University in 1971. Other colleges and universities created courses, programs, and centers on applied ethics, and where there are professors charting career paths, journals and textbooks are sure to follow. Some government agencies and foundations developed competitive funding programs to build capacity in ethics expertise and expand the ethics literature. As the 1980s unfolded there were similar developments elsewhere, especially in Western Europe.

New institutions were created and older ones more empowered. In the US, the thalidomide tragedy helped lead to greater authority for the Food and Drug Administration (FDA) to protect the country from dangerous medications. The Occupational Safety and Health Administration (OSHA) was created in response to decades of complaints about dangerous working conditions in many industries. The syphilis study revelations in 1972 led to a blue-ribbon committee appointed by the US government. Institutional Review Boards (IRBs, called Research Ethics Committees or RECs in other parts of the world), and Institutional Animal Care and Use Committees (IACUCs), were at first voluntary and then gradually came to be required under federal law. The new bioethics also had a way of drawing in those interested in high philosophical questions like the implications of advances in genetics and brain science for understanding the nature and future of humanity, as well was those concerned about more concrete matters like the limits of medical care at the end of life and innovations in reproductive medicine. It seemed that, under the very broad rubric of bioethics, there was something for everybody to wonder and worry about.

What Is Bioethics?

One may think of bioethics as composed of four subfields: the ethics of experiments involving humans and nonhumans, ethics in the practice of medicine, ethics in basic science, and the ethics of public health. My account will begin where discourse in medical ethics has long been anchored: in the conditions that justify human experiments. But it will lead from there to certain more general bioethical markers, like the proper relationship between doctors and patients, and finally to the settings of the laboratory and public health.

Taken only as an academic field, bioethics might seem to be of no particular significance in a discussion of the rules-based order and certainly of less moment for this purpose than fields like economics, political science, or the law. But in fact, bioethics is eminently representative of the rules-based order. There is a rich network of bioethics treaties and conventions, national regulatory requirements for research with human participants, credentials for research protection programs and in clinical ethics, and professional roles and organizations. At the global and intergovernmental level, the International Health Regulations (IHR) are legally binding rules for coordinated surveillance and response to public health risks. They apply to the World Health Organization (WHO) including, at least up to the second Trump administration, 196 states parties. Rooted in nineteenth century "sanitary conferences" created to respond to the spread of infectious diseases in Europe, IHR requirements include informed consent, data privacy, and nondiscrimination. The World Medical Association (WMA) pronounces on ethical issues in the medical profession from a global standpoint. Although its history dates back to the 1860s, the International Committee of the Red Cross (ICRC) received an international legal mandate through the 1949 Geneva Conventions. And there are treaty organizations that develop conventions against biological and chemical weapons on humanistic grounds or create platforms for war crimes tribunals that may include violations of medical ethics. Much of bioethics operates within the realm of soft law—guidelines and norms that lack legal force but are nevertheless widely, and at least formally, respected.

Complicating the picture still more, a broad scattering of careers, professions, and activities could come under the rubric of bioethics. Unlike traditional medical ethics, which was largely confined to the professional culture of medicine and medical education, bioethics has also become broadened, formalized, and institutionalized (some would say co-opted) in ways that reach far beyond the older medical ethics. There are jobs for bioethics professors in higher education and (mainly in the US) for clinical ethicists in hospitals—many of whom are trained as philosophers, not medical doctors, a development that would been little short of unthinkable for previous generations of physicians. In a number of countries, there are vigorous professional organizations, such as the American Society for Bioethics and Humanities (ASBH); the International Association of Bioethics (IAB), an international professional organization; as well as other more specialized groups such as the International Neuroethics Society (INS), Public Responsibility in Medicine and Research known (PRIM&R), and others; national bioethics commissions like the Nuffield Council in the UK; and an International Bioethics Committee (IBC) under the United Nations Economic, Social and Cultural Organization (UNESCO).

I was a member of the IBC for eight years, from 2010 to 2018. The invitation came out of the blue; like most Americans, even American bioethicists, I had never heard of the IBC, or if I had I had paid it no attention. As I quickly learned, the thirty-six-member committee is designed to be composed of experts, not representatives of their respective countries. I accepted out of curiosity and nostalgia. Some of my favorite childhood memories are of trips to Europe with my parents, noted psychotherapists, beginning in 1957. Every two years or so they embarked on lecture tours that, all told, took us from London to Moscow and many points in between. The IBC appealed to my love of global travel and making connections with colleagues far and wide. As a member, I learned the rituals and language of global diplomacy (states parties, low- and middle-income countries [LMICs], etc.), experienced the frustrating limitations of what UN entities are permitted to say and not to say (e.g., no singling out member states for praise or blame in official reports), and the provincialism of American bioethics. I was struck that many low-income countries, lacking the resources to assess the implications of new technologies, take the guidance of the IBC and other such UN entities as crucial for their policymaking.

My experience on the IBC also reinforced my sense that well beyond the clinical setting bioethics is a presence in the conduct of basic science, in global health, in public health policy and regulation, and national security organizations. Drug and device companies seek ethics consultation, and some have their own bioethics advisory panels. Considering all the members of research ethics committees and hospital ethics committees, ethics professors, social scientists, legal advisors, health policy experts, and health care professionals who have occasion to write about bioethics or work in its institutions, the numbers of participants in bioethics globally surely reaches into the many thousands. Are these people "ethics experts?" It's hard to say, but considering the demand, especially in a secularizing world, either there are ethics experts or some powerful sectors of society very much want there to be.

A World in Transition

I view the history of medical ethics and modern bioethics against a background of a changing global order. The transition to modern bioethics is a creature of the order alleged to have characterized the world since World War II. So, it is to be expected that another shift will challenge bioethics to respond. This book is not an attempt at forecast. Despite repeated attempts over the last couple of hundred years, the outcomes of large historical forces have resisted prediction, however much insight they might provide along the way. And there are wild cards like the effects of powerful personalities; Hegel's apocryphal observation of Napoleon as the "world spirit on horseback" comes to mind. But in a somewhat more modest vein, experience suggests that events on the scale of the pandemic; the wars in Ukraine, the Middle East, and Sudan; and a broad loss of confidence in neoliberalism and globalization have propelled the world into an uncertain new era. These uncertainties are multiplied by conditions that include immigration crises, Brexit, the rise of China, the early consequences of climate change, and, perhaps most especially, the emergence of right-wing governments that explicitly challenge the post–World War II consensus. What is clear is that all sorts of institutional arrangements and the norms that undergird them will not be unaffected by these struggles about international relations, including those of bioethics.

Readers in search of a comprehensive introduction to bioethics in the United States should look elsewhere.[7] Nor do I aim to provide a normative analysis of substantive bioethics issues within the rubric of an embracing theoretical framework; that is ground that is ably covered by a number of philosophers.[8] Rather, my project is an historical and sociological assessment of bioethics as a creature of the theorized rules-based international order. Some readers accustomed to reading about bioethics will be unfamiliar with what is in effect an ongoing conversation in political science and the history of ideas, while others will be unfamiliar with a field commonly associated with the realm of health care and medical science. I hope that each readership will find the integration I'm trying to achieve illuminating enough to be worth the trouble, for, according to the argument of this book, bioethics matters well beyond its native domain.

Bioethics captures and relies upon many of the assumptions that have characterized the aspirations of the post–World War II world, like universal human equality and ordered liberty under democratic governance. One might even say that bioethics is a paradigmatic expression of that postwar liberal project. Substantial changes in a global system in which these assumptions are not sustainable would be a very different world indeed. Only days before the final draft of my manuscript was due at the publisher, President Trump appeared to make good a threat made in his first term, that the United States would withdraw from the World Health Organization (WHO), rather than to reform alleged shortcomings. The mere expression of such a policy by a WHO founder and major donor indicates a drastically altered rules-based order.

Finally, I write as an American, not as a bioethicist with a view from nowhere, whatever that would mean. In every case when I have offered a generalization about the field, I have tried to be self-critical about an excessively American account. In a book with the kinds of aspirations I have here—to chart a history of bioethics as a global endeavor—that would be more than embarrassing; at the ambitious scale of analysis I aim for in this book that is perhaps unavoidable. Like any author I hope for a generous reader, such that—disagreements about details notwithstanding—the time is right to appreciate the bioethics conversation on a global stage.

A Note to the Reader

In the weeks before my April 2025 submission deadline, two seemingly unyielding forces rumbled ominously through the rules-based order: artificial intelligence and Donald J. Trump. The applications of AI are already reaching into every dimension of the human experience, including existential risks. From autonomous lethal weapons to novel toxins to chatbot psychotherapists, all implicate bioethics. Often built on mental models inspired by neuroscience, few technologies have been more in need of some sort of ordered, international consensus than AI. In principle the challenge is widely recognized; hardly a week seems to pass without a new declaration of proposed AI governance, but none can be effective without a global system that creates incentives for compliance. In a lawless 21st century imperial order artificial intelligence introduces a wild card that could further destabilize the traditional dynamics of nation-states.

The Trump administration appears prepared to ignore the international conventions upholding state sovereignty that the United States helped established after World War II. The conditions that built American health, wealth and global power are under assault. Thousands of federal workers key to the American research enterprise have arbitrarily lost their jobs, many stranded in foreign countries. In medical science "stop-work" orders for clinical trials and their legal challenges have left participants in medical research studies, some receiving long-term prescription therapy, others with study devices still in their bodies, in limbo. These measures do more than jeopardize a core value of medicine since antiquity, the non-abandonment of vulnerable people under one's care; they exploit a willingness to be part of clinical studies for the common good.

Every day is the opportunity to make further edits, but eventually lines must be drawn, page proofs must be finalized, and a book must be published.

History waits for no author.

Acknowledgments

For an author, acknowledgments can be the scariest part: Who will I have insufficiently recognized? That anxiety-laden question is linked to a prior one: When did I start writing this book? In this case the timing is especially ambiguous.

I trace my fascination with geopolitics to 1961 when I picked up a weighty tome that had recently arrived in my parents' library called *On Thermonuclear War,* in which the RAND strategist Herman Kahn rather ghoulishly set out the ways that nuclear weapons should figure into national security planning. A year later, the radius of the fireball and secondary blast effects of an atomic bomb targeting Manhattan as ground zero, sixty miles south of my Hudson Valley hometown, must have been on my mind. The Cuban Missile Crisis caused me to pay close attention to the sky when I walked home from elementary school. Let me therefore recognize the many relevant conversations about atomic-era great power politics that I had with teachers and friends over the ensuing fifty years, with much gratitude to those living and dead.

In the mid-1990s, my career in bioethics intersected with work for a presidential advisory committee on human radiation experiments. Chaired by Ruth Faden and directed by Dan Guttman, I came to appreciate the ways that bioethics was nested in the history of international relations. That revelation led to a series of writings about human experiments, biosecurity, neuroscience, genetics, and the history of science. This book is in some ways a culmination of all those pages.

I might not have attempted to write this book had it not been for an invitation from John Ikenberry in 2018 to be a guest for a meeting of a group of experts on international relations. For that session I was asked to prepare a several-page paper. The exact charge I no longer remember, but the result was a concise statement about the place of bioethics in the rules-based international order. I was heartened by the favorable response to the piece on the part of people who knew a lot about international relations but little or nothing about bioethics. Coming away from that session I had the feeling that I might have the basis of a worthwhile project. I am grateful to Ikenberry and the other members of the group in attendance that day.

I owe much bowing and scraping to Jonathan Marks, who read a previous draft that, as he made me understand, turned out to be far from near final in terms of its treatment of international human rights law. My postdoctoral fellow, Jonathan LoTempio also did a thorough read and made some important observations, tutoring me in data science and ethics.

Over the years, my friend Ulf Schmidt has opened my eyes to the neglected corners of medical ethics in twentieth century Europe; readers will note my deep debt to his important work.

Among many others who have influenced my thinking during the course of this project are Ron Bailey, Yelena Biberman, Alta Charo, Johannes van Delden, John Evans, Nick Evans, Tom Gage, Hank Greely, Amy Gutmann, Steve Joffe, Jonathan Kimmelman, Sergio Litewka, Paul Lombardo, Matthew McCoy, Eric Meslin, Jing-Bao Nie, Judit Sandor, Harald Schmidt, Nataliya Shok, Dominic Sisti, David Welch, Steve Xenakis, and Rebecca Yarvitz. Karolina Zhukoff translated some Russian texts. Kelly Hills did a wonderful job shaping my rough prior draft into a format acceptable to the MIT Press. More than that, she identified substantive issues that required my attention. No doubt I should have listened to all of them more; in any case they bear no responsibility for the results. Thanks as well to three anonymous reviewers, one of whom called this project a "geopolitical genealogy of morality," a thoughtful characterization that I much appreciated.

I am deeply grateful to the University of Pennsylvania and to my colleagues in the Department of Medical Ethics and Health Policy and in the Department of History and Sociology of Science. I hope that I have at least somewhat deserved Penn's generous support of my work.

My friendship with Art Caplan, the editor of this series, began fifty years before I finished this book. Within a few years of our first meeting at a graduate student party, and quite coincidentally, we had become colleagues as well. Suffice it to say we've seen a lot of life together in the ensuing decades. Everyone in every field should have a cherished friendship like this.

Thanks also to my editor at the MIT Press, Phil Laughlin, for seeing this project through to completion even as last-minute adjustments were required in light of the transformative events of 2025.

Our grandson Jude Liam Moreno made his entrance amid the completion of the penultimate draft. I find it impossible to imagine how impoverished my life would be without Leslye, Jarrett, Jillian, Reiko, Brandon, and now little JL.

1 The Long Peace

The bioethicist and historian Robert Baker has written that the liberal order that emerged after World War II shaped bioethics "as an ethics of covenants and conventions grounded in human rights."[1] If Baker is right, as I believe he is, then geopolitical change poses a challenge to bioethics generally, both in theory and in practice, especially regarding the status of human rights. In order to reconstruct the rise of bioethics within the post–World War II rules-based order, it is important first to understand the historical foundations of the conception of the modern nation-state, supposed in liberal democracies to be the guarantor of human rights. Without that foundation the idea that there are such national entities that can and do interact and establish norms for those interactions, including international bioethics norms, is harder to appreciate.

The International Order

The rules-based order of which bioethics is a creature can be said to have three aspects: finance, security, and human rights. Modern bioethics can be viewed mainly as part of the human rights aspect of the rules-based order. However, certain values that define modern bioethics also have a secondary but essential role in assumptions in the financial and security dimensions. For example, the economic sanctions imposed on the Russian Federation following the Russian annexation of Crimea in 2014 were founded in part on the view that Russia was in violation of the right to popular self-determination within the boundaries of a sovereign state. The security dimension of the liberal order includes conventions about weapons thought to be inhumane. Thus, certain understandings about values of human agency and dignity, although especially obvious in bioethics as part of the human rights dimension, are pervasive in the rules-based order. If there is a case to be made along these lines, one that I will attempt to make in this book, that would vastly enrich our understanding of the place of bioethics as the world has been shaped in the last seventy years. It will also raise significant and perhaps unsettling questions about the future of bioethics in a new global order.

The rules-based order stipulates that international relations are to be organized according to principles of open markets, liberal democracy, and multilateral organizations. The connection to liberal democracy is especially important for bioethics and associates it to one of bioethics' core values, variously expressible as personal autonomy or self-determination, a Kantian notion rooted in the special significance of moral agency and its implications for the universality of human rights. The various flavors of consent in bioethics—including voluntary, informed, implied, and inferred—operationalize respect for autonomy. As I will explain, this critical norm of the rules-based order with which we are now familiar took some time to become so prominent. The history of modern bioethics is a central part of that story.

None of the conventions and standards I will describe were formulated simply from the milk of human kindness. The motivations were not grounded in goodness, or a sense of charity, nor did they emerge from polite conversation following a garden party. Their foundations owe more to Hobbes than to Rousseau, a series of reactions to periods of massive barbarity, cruelty, and suffering. Nor did they spring from a collective insight into the imperative following a philosophy seminar conversation in response to a professor's blackboard scrawl, "Why be moral?" Although often ambiguously expressed, fitfully embraced, and rarely enforced, the alternative had been experienced and was unacceptable.

In the words of one of its most prominent theorists and advocates, G. John Ikenberry, "Liberal internationalism can be understood as a cluster of ideas on how to think about and act in the world, which emerged out of the Enlightenment and the Western liberal democratic experience."[2] The rules-based order was motivated by the trauma of two world wars (or, some would say, one long war with a misleading interregnum), that wreaked horrific damage in both material and moral senses. Amid the annihilation of tens of millions of people and incalculable financial loss, one of the victorious countries remained largely unscathed and, for a time, with largely unrivaled power. In a passage that may be read with a certain wistfulness after 2025, Ikenberry continues:

> In the aftermath of World War II, the United States and its partners built a new type of international, organized around open trade, cooperative security, multilateralism, democratic solidarity, and American leadership. Regional and global institutions were established to facilitate cooperation, enshrine shared norms, and bind societies together. Western Europe overcame centuries of division to launch a project of integration and political union. West Germany and Japan reinvented themselves as "civilian" powers and became stakeholders in the postwar liberal order. . . . When the Cold War ended and the Soviet Union collapsed, this Western liberal order expanded outward across the globe.[3]

Theorists of the rules-based order cite its intellectual roots in the Enlightenment, along with the very idea of modernity, that the world need not be static and respond only to nonhuman forces, and that human intervention can change it. Therefore, progressive improvement in the human condition, at least in its material aspects, can be accomplished by human decision making. Neither is moral improvement guaranteed but human beings are capable both of appreciating institutions as the setting in which society operates and of managing them. Nor are different parts of the world ultimately isolated from one another; ours is a globally interconnected system that moves in the direction of a common future. In various ways thinkers like Condorcet, Hume, Kant, Montesquieu, and Adam Smith, as well as America's own Benjamin Franklin, contributed to the idea of modernity as inherently a product of human agency. The idea of modernity was the seedling of the idea of a rules-based order.

State Sovereignty

Sovereignty is a prerequisite for a global order, a condition for nation-states' entrance into agreements. Historians generally attribute the origins of the modern concept of international relations to the treaties that made up the Peace of Westphalia in 1648. These conventions were direct responses to the calamitous Thirty Years' War that had taken millions of lives and untold treasure, though the treaties themselves took years to negotiate with many parties participating at various times. The war was in part a sectarian conflict between Catholic and Protestant powers, especially the Habsburgs on one side and Sweden on the other, though Catholic France was allied with the Protestants because of bitter opposition to the Habsburgs, a rivalry that continued for many years after the treaties. The war was also a struggle over the Roman Church's influence over Europe. Despite furious papal objections, under the treaties, states were allowed to define their own official religions. National territorial boundaries were set and considered to be inviolable.

Historians disagree about whether these elements amounted to state sovereignty, and the term does not appear in the treaties, nor does it exist in the Latin vocabulary of the texts. Agreements that presupposed sovereignty had already been made between "estates," high-ranking classes with political power. However, in the words of one historian, "The wording used to effect the transfers [of formerly disputed people, places, and things] appears to support the idea of a new and more sharply defined notion of sovereignty" in the Peace of Westphalia.[4] Another writer sees Westphalia as more definitive:

> The Peace of Westphalia of 1648 famously recognized the secular rights of German princelings above the religious claims of the Papacy, legitimating no external power beyond that of the sovereign; it was this formal recognition of the principle of territorial sovereignty which henceforth became the basis of relations between states. . . . Without international law, the regulation of inter-state relations could not extend beyond voluntary agreements between the sovereign states—strategic alliances, aimed at preserving local interests and maintaining a relatively stable balance of power.[5]

In the final analysis, the Peace of Westphalia didn't substantively resolve conflicts among the powers so much as constitute an agreement to keep hands off the others' internal affairs. Its critics note that enforcement was left to the discretion of the powerful who were still at liberty to interfere in the affairs of the weaker: "Under the Westphalian system, then, superior force was the guarantor of effective sovereignty."[6] But, according to Westphalia's admirers, it did at least suggest that disagreements had to be resolved by treaty between sovereign states, precisely because of the principle of noninterference, though often honored only in the breach. Westphalia did more or less have that result, with far fewer armed conflicts in Europe over the next couple of hundred years. However, colonial domination by major powers in Africa and Asia was unconstrained.

The Peace of Westphalia might strike a twenty-first-century reader as an obscure reference, and it certainly falls far short of a rules-based order. But it is far from an irrelevant curiosity and, in principle at least, it was a breakthrough. Westphalia forms part of the basis of the United Nations charter and of the international order itself. In Ikenberry's words,

> Westphalian internationalism is cooperation between sovereign nation states. It is manifest in foundational agreements about the norms and institutions of the sovereign state system and in more specific functional agreements in areas like security, economics, and the organization of the global commons. Various strands of nineteenth-century liberal internationalism built on and affiliated themselves with both of these foundational types of order. In the twentieth century, as European empires weakened and broke apart and the United States rose to global dominance, liberal internationalism increasingly tilted in the direction of building its projects on the foundation of an expanding Westphalian state system.[7]

In 1815, the Congress of Vienna implicitly recognized the fundamental importance of sovereignty when, following the Napoleonic Wars, it reordered European territorial boundaries in hopes of a new balance of power that could keep the peace. Immanuel Kant is typical of Enlightenment philosophers in presupposing state sovereignty as a condition for the achievement of "perpetual peace" within and among nations, ideally achieving a global federation of states.[8]

Within a few years the Concert of Europe imposed a great power agreement over lesser European states to settle claims after the Napoleonic Wars. Subsequent nineteenth-century "concerts" also reflected nation-state cooperation, at least among the major powers who agreed to respect territoriality, as well as creating an idea of Europe.

In hardening the concept of physical boundaries between states, Westphalia did create a platform for international order but worked against international integration. For example, under the liberal international order human rights claims are said to be universal, but the Westphalian system doesn't allow for cross-border enforcement of such claims. Besides, it wasn't until the mid-twentieth century and the postimperial era that norms of political equality and universal human rights became part of the international order. Like the rules-based order itself, nation-states must agree to be governed by them. In one sense a threat to the liberal international order is a threat to the modern consensus about the universality of human rights.

In another sense, state sovereignty must itself be constrained by human rights. As John Rawls observed, "Since World War II international law has become stricter [for] it tends to limit a state's right to wage war to instances of self-defense . . . and it also tends to restrict a state's right to internal sovereignty. The role of human rights connects most obviously with the latter change as part of the effort to provide a suitable definition of, and limits on, a government's internal sovereignty."[9] In 2005, a specific form of Rawls's observation was unanimously adopted by all heads of state and government at the UN World Summit. The international norm known as the "responsibility to protect" (R2P) was a response to atrocities in Rwanda and the former Yugoslavia.[10] State sovereignty and its limitations under a rules-based order thus have practical implications for the universality of bioethics as a set of norms about human rights in the setting of medicine and the life sciences.

Creating Modernity

As the nineteenth century unfolded, the rapidity of change was obvious in much of the world, particularly in connection with technological innovations like the railroad and the telegraph, built on the industrial revolution and the spread of capitalism. These developments affirmed and fitfully advanced the Enlightenment notion of a global order. Or, one might say it was a globalized disorder that was itself evidence of modernity. Weapons became far more lethal, forcing gradual change in traditional ideas of the nature of warfare, of both tactics and strategy, but implementation of new knowledge was slowed by institutions that resisted reform. In medicine, technologies like the stethoscope and evidence about the importance of hygiene required generational time to adopt. In Europe the imperial system into which international relations were organized for hundreds of years was upended with the first fifty calamitous years of the twentieth century. During the intervening centuries, imperialism came to be associated with raw might such that stability among the dominant states was only achieved by a balance of power. Thus the limits of Westphalia were laid bare. The balance of power broke down with the outset of World War I through a series of distantly related events that exposed the underlying instability of the Westphalian system, setting up a difficult transition from empires to nation-states in which the Enlightenment notion of global integration could be realized. This, at least, was the liberal dream. At the 1919 Paris Peace Conference, Woodrow Wilson declared a principle of self-determination for the whole world but, like other elements of Wilson's aggressive internationalist ambitions, the major powers' colonial territories were left out of that principle until after World War II.

During the nineteenth century, the imperial and the Westphalian orders were in a period of mutual adjustment. Each was internationalist in its own way, so that the liberal international order of the early twentieth century included remnants of both. The old empires, such as those of Great Britain and France, struggled to survive through the world wars while the Treaty of Versailles (the World War I peace agreement) presupposed state sovereignty. So did the League of Nations, which was intended to provide a platform for forward-looking international cooperation built on international law. It was dismayingly clear that mercantilist trade policies, against a background of excessive nationalism, hadn't sufficiently tied nations together to avoid the Great War. (One thinks here of pandemic vaccine protectionism.) Something more was needed. Implicit in the aims of the League's founders was that the new order would provide an environment in which liberal democracies could flourish, a multilateralism with popular consent, so that the League itself would be a union of liberal democratic states.

In the end it didn't quite work out that way. Wilson's failure to reach a deal with the Senate to ratify League membership and the inherent flaws of Versailles gave the lie to the most ambitious plans for a new order. There were also divisions among progressives. Intellectuals like John Dewey, who saw the war's end as an opportunity for global reform, campaigned for an international arbitration mechanism and the outlawry of war, a movement Wilson didn't embrace. But there were substantial efforts. After the war, France sought alliances to prevent future German aggression, beginning with a series of naval conferences and culminating in an agreement that outlawed war and emphasized peaceful resolution of differences. Building on Andrew Carnegie's Endowment for International Peace, the Kellogg-Briand Pact of 1928, named for the American secretary of state and his French counterpart, declared war, or crimes against peace, illegal, forming the basis for the prosecution of war crimes in Germany and Japan following World War II.[11] On this occasion the US did sign this "self-defense" pact along with dozens of other countries. The pact was toothless and quickly collapsed. Considering how the rest of the century unfolded, Kellogg-Briand clearly failed in its primary mission, but it did draw a line between war and peace and established the norm that territory should not be acquired by naked conquest. Quite nearly all international norms since Kellogg-Briand are founded on its principles, including the UN charter.[12]

Wilson's internationalism concerned itself mainly with collective security and free trade, not with radical reforms about human rights or welfare. The cold war historian John Lewis Gaddis argued that "the Versailles Treaty did come about as close as any in modern history to incorporating within itself the principles of justice."[13] That is faint praise; there was no room in Wilson's vision for dark-skinned, colonized peoples. Even with that narrow focus it appeared that despite the horrors of the Great War, illiberal and autocratic European interests weren't quite ready for modernity. The old regime was teetering on the brink. Marxist-Leninist ideology and the Bolshevik Revolution emerged as new challenges to Western ideas of internationalism, which notably paid little attention to Asia and Africa. The League's Permanent Court of International Justice had a strong start hearing cases and clarifying international law, but its authority faded as the 1930s progressed. Amid a worldwide depression, and the rise of fascism and totalitarianism, there was a sense of lost opportunity. A debate erupted about the failure of capitalism. Was the villain the modern state that undermined freedom, as Friedrich von Hayek argued? Or, as argued by John Maynard Keynes, was more state spending the answer? Despite their differences, Hayek and Keynes were among those who were concerned with identifying the political, economic, and social conditions under which human beings could flourish in democracies. This was a far grander problématic than the previous era of thinking, one that appreciated the entanglement of national and international dynamics. Now, with burgeoning new technologies forcing a new modernity, another horrid world war created another opportunity to move decisively beyond the nineteenth-century imperial system.

The new internationalism was being designed by the Allies even before World War II ended. Wilson's ideas about collective security needed to be put in a frame adapted to newly complex and dangerous conditions. Domestically, Franklin Roosevelt enlarged what he called "liberalism," abandoning his cousin Theodore's "progressivism" and studiously avoiding "socialism," thus precipitating a confusion about the meaning of the word—classical liberalism was associated with limited government—that continues to this day. For Roosevelt, a liberal was someone who was committed to governmental assurance of security and welfare, principles embodied in his New Deal. This formulation implied a vastly enlarged government, not only in terms of bureaucracy but also in its mission of assuring social justice. (In economics, Keynes—who virtually created the field of macroeconomics—thus prevailed over Hayek, despite the latter's reputational revival in later twentieth-century conservative American politics.)

The Atlantic Charter, signed by President Franklin Roosevelt and Prime Minister Winston Churchill in 1941, is often taken be the first clear statement of the new rules-based order. It set out the two countries' goals for the postwar world, specifying economic cooperation, social security, and other commitments, like national territorial integrity, freedom of the seas, and popular self-determination. Just a few months later, the attack on Pearl Harbor undermined lingering American isolationism. As it was hard to see how the alliances formed during the war would go away, some form of internationalism as an acknowledgment of economic interdependence, at the very least, was unavoidable. The Bretton Woods Agreement in 1944 created the International Monetary Fund and the International Bank for Reconstruction and Development, among the first examples of the multilateral institutions. The United Nations came to embrace various entities like the International Court of Justice (not to be confused with the International Criminal Court), the General Agreement on Tariffs and Trade, the Food and Agriculture Organization (FAO), the International Atomic Energy Agency, the World Health Organization (WHO), and later the Organization for Economic Cooperation and Development (OECD) and the Codex for food safety. The spirit of the time was typified in the attitude of leading statesmen like Dean Acheson, who believed that "peace—and 'free enterprise'—depended on a general expansion of worldwide production, trade, and consumption. The IMF and World Bank would help this come about, just as the United Nations was intended to help order political relations."[14]

One must take care not to romanticize the institutions created in the wake of World War II. In particular, Bretton Woods fell far short of what Keynes envisioned as the most important goal of the new order: to stabilize the international monetary system in order to prevent future financial crises like those that, he believed, led to Fascism and Nazism. His proposal called for a new international bank to regulate the global supply of money and flow of currency for trade, with a new reserve currency called the bancor. Keynes' biographer is blunt about the reason that the Roosevelt administration, though friendly to Keynesianism in the shape of the New Deal, rejected his plan, because "in truth the U.S. government simply had no interest in creating an international order that would diminish American power."[15] The reserve currency would be the American dollar, creating leverage for the United States that remains in place well into the twenty-first century, despite the relative decline of American control of the international monetary system and various financial shocks and crises over the decades. One of these was the notorious "Nixon shock" in 1971, when the Nixon administration, fearing that the threat of inflation would impact his reelection, imposed wage and price controls and eliminated the US dollar's convertibility to gold, in essence ending the Bretton Woods system. There have been calls for a new Bretton Woods–style agreement to address a drastically different global economy.

For all the American self-interest enshrined at Bretton Woods, there was nonetheless what Ikenberry calls a "nascent universalism" in the new liberal international order. A shrinking world made universal equality a practical requirement, however grudgingly. Even "the people of Java are getting to be our neighbors," Roosevelt said to Harry Hopkins, his closest advisor.[16] This offhand remark belied that it was decidedly Western and American values and interests that were presupposed in all the planning for the new "liberal hegemonic" order, to use Ikenberry's phrase: "The core impulse that drove the United States in its postwar order building was its recognition that, given its unrivaled power, it had an opportunity to shape the international environment and set the rules in which it and other states operated—and it could do this with an eye to the long term."[17] Nonetheless, it is notable that international security in the post–World

War II order would have to involve not only physical but also normative dimensions. That meant that the people of Java would indeed have their say.

The Realists

The new liberal internationalists were not altruists. In the words of Gaddis, the new order "was based not upon the dictates of morality and justice but rather upon an arbitrary and strikingly artificial division of the world into spheres of influence. . . ." The internationalists saw themselves as responsive to a new set of facts, and if the response happened to create and sustain certain advantages for the United States and its liberal democratic allies in the postwar world, well, so much the better. In 1986, Gaddis expressed surprise that the post-World War II system of international relations "has now survived twice as long as the far more carefully considered World War I settlement. . . ."[18] As it turned out what Gaddis called "the long peace" was even more durable than that, lasting 80 years.

But what if the very idea of the rules-based order is an illusion, and a dangerous one at that? The political scientist John Mearsheimer, considered a leader of realism in international affairs, views the rules-based order as a myth propagated by modern liberals that has provided a convenient rationale for a few Western countries, the collective victors in World War II and the Cold War, to exercise their power. Mearsheimer also uses the term "liberal hegemony," but he defines it as "an ambitious strategy in which a state aims to turn as many countries as possible into liberal democracies like itself while also promoting an open international economy and building international institutions." The problem with this view, according to Mearsheimer, is that national self-interest must ultimately prevail, regardless of the language of liberalism: "Great powers are rarely in a position to pursue a full-scale liberal foreign policy. As long as two or more of them exist on the planet, they have little choice but to pay close attention to their position in the global balance of power and act according to the dictates of realism." Worse, he argues, in practice, liberal hegemony spawns resentment among lesser powers. Mearsheimer is careful to distinguish his views about foreign policy liberalism from liberalism in domestic government, creating the illiberal backlash: "States that pursue ambitious liberal foreign policies, as the United States has done in recent years, end up making the world less peaceful."[19]

For realists a prime example of liberal overreach in recent history is the invasion of Iraq in 2003. The "slam dunk" evidence of Iraqi biological weapons (despite the reporting of international organizations) turned out to be at best a horrific error and at worst fraudulent. Nor did Iraqis have any involvement in the 9/11 attacks. As a candidate, George W. Bush rejected "nation building," a view that led to a misbegotten reliance on a fantasized Iraqi polity. The president's advisors imagined an opportunity to create conditions for a liberal democratic state in Iraq, as well as a bulwark against Iran, with minimal Western guidance. Despite the differences in ideology, to Mearsheimer these "neoconservatives" make the same classical liberal mistake about human nature. "In practice," Mearsheimer writes, "the vast majority of people around the globe do not care greatly about the rights of individuals in other countries. They are much more concerned about their fellow citizens' rights, and even that commitment has limits. Liberalism oversells the importance of individual rights."[20] Mearsheimer would surely dispute Roosevelt's comment about the "people of Java" as having any relevance to the American people, much less the people of Iraq.

Perhaps Mearsheimer is right that the liberal international order is illusory, a product of an overheated left-liberal imagination. If that is the case it was a most prolific illusion, a hot dream of a global commons. A short list of the products of postwar cooperation would include the United Nations with its various functions, organizations, and conferences; the General Agreement on Tariffs and Trade; the World Trade Organization; the International Monetary Fund and World Bank; the Export-Import Bank; arms control and nonproliferation treaties; outer space and law of sea treaties; air travel and telecommunication regimes; and regimes to combat ocean pollution, ozone depletion, and on and on. Yet in an important sense, whether Ikenberry, the liberal internationalist, or Mearsheimer, the conservative internationalist, is right is a red herring because, whether true or not, the rules-based international order is the story the world has told itself. In either case, then, the world needs to be prepared for a drastic turn in history, as do the professors and practitioners of bioethics. That is because, as a practical matter, bioethics has a role in the foundations of the supposed rules-based order and has a stake in its future. Either the foundations of bioethics are under the greatest threat since World War II or those foundations have always been illusory. In either case ideas have power: The very idea of the rules-based order predicated on universal human rights has had practical effects.

Human Rights in the New Order

Mearsheimer's skeptical remarks about individual rights and self-determination appear in the context of the use of national power in both armed and unarmed conflict. Nonetheless, they also set up an interesting question about bioethics without liberal internationalist assumptions. However, no country is likely to go to war to assert the rights of human research participants. For example, the regulatory and professional institutions intended to protect their rights as universal are without question part of the rules-based order. The founder of the concept that nations should be bound in their relations by the rule of law was also the first to argue that rights be attached to persons rather than only to objects. This was another Enlightenment-era thinker, the Dutch scholar Hugo Grotius. Keen to find a way to limit the calamities of war, Grotius developed the idea of justice among nations based on the writings of ancient Greek and Roman philosophers. What he called "natural law" is accessible to all rational creatures. Concerning conflicts between nations, Grotius distinguished between wars of aggression and wars of defense. If a war must be fought, it should be governed by certain rules, including the humane treatment of prisoners.[21] In this sense, there is an implicit idea of universality in Grotius's arguments, that at least prisoners of war should be treated the same, regardless of their origin.

The idea of human rights "declares that every individual has legitimate claims upon his or her own society for certain freedoms and benefits."[22] The American Declaration of Independence (1776) cited inalienable rights to "Life, Liberty, and the pursuit of Happiness," and the French Republic (1789) declared the "natural, inalienable, and sacred rights of man." The two world wars forced a reconsideration of human rights. In the post–World War I treaties, the concept of human rights was attached to membership in ethnic and religious groups as constituents of sovereign states. But that system failed to consider that some groups, like Jews and Roma, were not associated with distinct sovereign territories, a limitation that was all too clear after the Holocaust. Nor were minorities of other sorts considered, like people who would now be understood to be part of the LGBTQIA+ community or disabled. Thus, there was a distinct shift in the World War II Allies' postwar planning from the assurance of group rights to that of individual rights, as reflected in the language of the founding documents of the United Nations. Its 1942 Declaration asserted the "preservation of human rights." The preamble to the UN Charter (1945) "reaffirm[s] faith in fundamental human rights, in the dignity and worth of the human person, in the equal rights of men and women and of nations large and small." (This last phrase effectively renounces the old imperial system in which the great powers fell into their own category.) Articles 55 and 56 of the UN Charter commit member nations to "universal respect for, and observance of, human rights and fundamental freedoms for all without distinction as to race, sex, language, or religion." Article One of the Universal Declaration of Human Rights (1948), considered to be the cornerstone of modern international law, states that "all human beings are born free and equal in dignity and rights. They are endowed with reason and conscience and should act towards one another in a spirit of brotherhood."

Linked to the modern understanding of human rights as associated with persons, bioethical values are written into the fabric of the post–World War II rules-based order. Citing again only declarations within the UN system (all nonbinding but considered important elements of international law), in 1997 the United Nations Economic, Social and Cultural Organization. General Conference adopted the Universal Declaration on the Human Genome and Human Rights, citing principles of respect for human dignity, noncommercialization, benefit sharing, and the importance of scientific progress. Next came the International Declaration on Human Genetic Data (2003), a model for legal protection of human DNA. The Universal Declaration on Bioethics and Human Rights (UDBHR, 2006) states that "human rights constitute the tangible elements for achieving . . . freedom, and therefore, peace as progress, as transformation and as deepening in dignity, equality, and freedom of all human beings." The UDBHR alluded to a wide range of principles including personal autonomy, consent, privacy, equality, and benefit sharing. Two UNESCO entities, the International Bioethics Committee and the World Commission on the Ethics of Science and Technology, have published reports on such topics as nondiscrimination and nonstigmatization, medical ethics and refugees, and adaptation to climate change. The Council for International Organizations of Medical Societies (CIOMS), established by UNESCO and the World Health Organization, publishes specific ethics guidance documents on topics like drug promotion. In all this, bioethics is deeply embedded. The bioethicist and legal scholar Judit Sándor has observed that "as a result of the development of new human rights in the field of biomedicine and the development of normative bioethics, bioethics and bio(medical) law have mutually borrowed statements, norms, principles, and perhaps even methods from each other."[23] I will return to that important point later in this book.

In international relations, "soft laws," like these UNESCO declarations, include private or governmental recommendations, guidelines, nonbinding resolutions, declarations, and the like. They lack legal force as within, say, the Geneva or Hague conventions and are not buttressed by enforcement mechanisms but, in the words of one commentator, "they can trigger a spontaneous coalescence around a standard in a 'snowball effect' fashion."[24] Much of bioethics functions as a manifestation of soft law, expressed though what sociologists of knowledge call an "invisible college" of peers in an extended professional community. Sometimes the invisible college is not so invisible. An example is the InterAcademy Partnership (IAP), an umbrella organization of 140 international, national, and regional academies of science, engineering, and health professionals around the world, accounting for thirty thousand members. The IAP provides policy advice to governments and intergovernmental organizations. Part of its mission is the transmission of scientific standards to future generations of researchers. Its 2012 report argued that the values of scientific research are universal.[25]

To be sure, this professional cultural transmission is the softest of soft law, with standards that are often honored mainly in the breach. What is surprising is not that the practical efficacy of soft law varies, but that it even functions to begin with. How can such an informal system with no power to impose sanctions operate at all? One explanation is what has been called "network effects" or "network externalities," which appear "when the value of a product or service increases as the number of other agents using the same product or service grows, which in turn draws more users. Network externalities arise from the need for compatibility between standards."[26] Examples range from natural language between users to telephone networks, computer operating systems, time zones, and the sizes of screw threads.

Soft law institutions vary in network strength based on the participants' needs to understand how to render their behavior compatible with that of others. As a member of the International Bioethics Committee, I was struck that the IBC and its reports are hardly known in the United States (the global hegemon that sets the standards often doesn't ask what they are), but they are prominent as sources of expertise and policy guidance in much of the rest of the world. IBC reports are especially appreciated in low- and middle-income countries (LMICs) that want to ensure that they are in synchrony with the world order, appreciate knowing the rules of the road, and don't have a great deal of their own policy expertise. As expressed by a bioethicist with international experience, "More important than contributing to the adoption of normative instruments the IBC has played a leading role in promoting a broader conception of bioethics that is more appropriate to current processes of globalization."[27] The informal influence of the IBC may be even more significant than its policy statements. In that respect, my litany of UN declarations and organizations, complemented by other milestones I will discuss, is meant to identify a network of norms that are either foundational to bioethics, like the Universal Declaration of Human Rights, or specific to a substantial bioethics mission that has contributed to the development of an international framework of assumptions about public and private conduct.

The Law of Armed Conflict

In contrast with soft law, a salient example of binding international law amid the human rights dimension of the rules-based order is the modern law of armed conflict (LOAC) as embodied, for example, in the Geneva Conventions (1949). The basic principles of the LOAC, also known as international humanitarian law (IHL), are military necessity (a belligerent does what it is necessary for its military ends and avoids that which is unnecessary), humanity (forbidding the infliction of suffering, injury, or destruction unnecessary for legitimate military purposes), and chivalry (some forms of deception in wartime are tolerable but not perfidy or treachery).[28] To this list some would add the distinction between combatants and noncombatants and proportionality of response to an attack, referencing Article 51.3 of the Geneva Conventions.

With significant commonalities with principles of international human rights, these LOAC principles stand behind the more than fifty conventions regulating armed conflict, beginning with the 1856 Paris Declaration Respecting Maritime Law and culminating with the 2008 Convention on Cluster Munitions. The Geneva Conventions have been formally adopted by virtually every nation-state. Although both tacit and explicit agreements about humanity in warfare go back thousands of years, to the ancient Greek city-states (significantly excluding "barbarians"), the process leading up to the modern explicit Conventions can be traced back at least to 1864 and through to the turn-of-the-century Hague Conventions. The latter included the 1907 Convention on the War on Land and the 1980 convention on Certain Conventional Weapons, as well as other protocols, declarations, and statutes both within the Geneva and Hague frameworks and outside of them, too numerous to mention in a brief overview. Expressions of the LOAC can also be found in military manuals, national military criminal codes and case law, international criminal tribunals and conferences, and the Red Cross and Red Crescent. A primary function of the ICRC is that of educating both nation-state and non-state actors' armed forces about the LOAC.[29]

Despite the presence of "humanitarian" in its alternative name, the LOAC is harsh. It prohibits targeting noncombatants but not their deaths if that collateral damage is justifiable according to military necessity and is proportionate to operational goals. As a practical matter, no nation-state contemplating the need to protect itself would agree to prohibition of noncombatant deaths or injuries. Like any laws, the LOAC is subject to interpretation. It can be and often has been violated. Alleged violations have become especially prominent in the urban landscapes of twenty-first century conflicts, but this is not argument against the status of the LOAC as international law or of the interests that potential adversaries have in its preservation. (If laws that were commonly violated were thus rendered inoperative there would be no stop signs.) The LOAC with its many provisions is not only widely held and incorporated into the training of many major and minor national armed forces, to the extent that it is respected it creates some conceptual difficulty for realists who might expect it to be routinely flaunted. Consistent with the serious implications of these agreements, there are certain understandings about the application of the LOAC: They are generally the results of multilateral conferences with their own procedural rules; allow states to formulate reservations; could be followed by successive, clarifying treaties; are to be interpreted in good faith among the state parties; make provisions for relations with states that are not parties to a particular agreement; and provide rules for the termination of a treaty as the result of a breach but not, importantly, in a way that violates humanitarian conduct. There are also implicit or customary aspects of the LOAC. These are practices considered binding in all states if they are extensively, virtually, and uniformly occurring, and if there is a belief that a practice is required, prohibited, or allowed.[30]

The Geneva Declarations and the Universal Declaration of Human Rights were profoundly influenced by the revelations of Nazi crimes through trials at a number of sites throughout Germany and especially the most high-profile proceedings in Nuremberg. Among those crimes were gross violations of medical ethics, crimes that were found to violate the law of war despite arguments about their military purpose. It is to those events and their role in forming bioethics as a core project of the rules-based international order that I now turn.

2
Achieving the Nuremberg Code

Second perhaps only to the Hippocratic Oath, the Nuremberg Code is the most cited and most influential document in the history of medical ethics. But the Code's essential place in the concepts that comprise the postwar rules-based order is not widely appreciated. More than a set of ethics standards for human experiments, the Code is the specific application of a broader set of principles that guided the indictments and judgments in thirteen legal cases in Nuremberg after the war. Well beyond those tribunals, the "Nuremberg Principles," not to be confused with the Code, are widely recognized in international criminal and human rights law. They played a critical role in the establishment of the liberal international order. As I will explain in this chapter, the Nuremberg Principles themselves emerged from centuries of evolving standards about the behavior of nation-states during armed conflict. Thus, the ten-point Code produced by the judges stands within a network of concepts that extends far beyond the ethics of human experiments.

The failure of bioethicists to appreciate the role of bioethics in the rules-based order is partly a result of a lack of awareness of the Nuremberg Principles, as distinct from the Code. The Principles can be viewed as creating the context for the Code, and the Code in turn specifies the core values of the Principles in the setting of human experiments as a key practice of medical science. Significantly, after the trial of leading surviving figures of the Third Reich, the second Nuremberg trial focused on what was known among the organizers of the prosecution as the medical case, better known to posterity as the doctors' trial. Before examining the trial itself, however, it is useful to describe in some detail how laborious and frustrating was the path to the Nuremberg Principles and the International Military Tribunal (IMT) that prosecuted the first case. The Principles both form much of the basis of, and presuppose, the imperfect but consequential rules-based order, following a path that wends its way through great power politics and the aftermath of a war whose horrors were barely comprehensible then and seem still harder for subsequent generations to grasp.

War Crimes

The codification of war crimes is a cornerstone of the rules-based order. Signed by the Allies in 1945, the Nuremberg Principles created by the IMT codified three types of crimes: aggression or "crimes against peace"; war crimes or "violations of the laws and customs of war," including murders, "wanton destruction," and "devastation not justified by military necessity"; and crimes against humanity, "namely, murder, extermination, enslavement, deportation, and other inhumane acts committed against any civilian population, before or during the war; or persecutions on political, racial or religious grounds in execution of or in connection with any crime within the jurisdiction of the Tribunal, whether or not in violation of the domestic law of the country where perpetrated." The Nuremberg Principles also stipulate that heads of state are responsible for crimes under international law and that the defense of "superior orders" does not necessarily release a subordinate from moral choice: "The fact that a person acted pursuant to order of his Government or of a superior does not relieve him from responsibility under international law, provided a moral choice was in fact possible to him."[1] The latter Principle turned out to have especially important implications for the medical case.

The Nuremberg Principles were far from arbitrary or unprecedented, nor were they detached from the practicalities of organized violence between armies but emerged from long memories of its horrible consequences for combatants and noncombatants alike. International law has not emerged from human goodness, but from prudence in light of harsh experience. In his magisterial account of the Nuremberg trials, Army lawyer and associate trial counsel for the IMT prosecution Telford Taylor cautions against "the wholly mistaken notion that the Nuremberg Trials were the first 'laws of war.'"[2] Wars in feudal Europe were often fought by mercenaries whose reward was partly whatever they could plunder from invaded lands. Fighters were brutal and brutalized, disrupting villages and farms, and turning locals against them to create a threat in the rear, never a desirable situation for an aggressor force. Commanders were frustrated by poor discipline in the ranks. Just as the Thirty Years' War led to the Peace of Westphalia, which implied a measure of state and territorial sovereignty, it also impelled empires to create organized forces under a chain of command with supportive logistical arrangements for regular pay and supplies. Regulations were established with military police and courts-martial to enforce them. Noncombatants were off limits. Although these military regulations did not rise to the level of international law (or ultimately to the LOAC), they were intended to render armed conflict more humane, less chaotic, and less indiscriminately destructive.

It wasn't until the middle of the nineteenth century that elementary laws of war were codified rather than only customary. In 1863 the Union Army issued General Orders 100, "Instructions for the Government of Armies of the United States in the Field." Known as the Lieber Code for Prussian-born, reformist author Francis Lieber. The Code included rules about the treatment of prisoners and spies and a ban on poisons. The General Orders gave wide latitude to "military necessity," but condemned cruelty to civilians, prisoners, and wounded soldiers that did not advance legitimate war aims. Attempts were also made to bring order to field medicine. Modern arms caused dreadful injuries for which neither tactics nor medical support were prepared. There were similar experiences in Europe in the previous decade during the Crimean War and the Franco-Austrian conflict in Solferino. The immense suffering inspired the first Red Cross convention in Geneva in 1864 that set rules for the care of wounded soldiers.

Gradually, the LOAC circumscribed the application of military necessity through the recognition of war crimes. A series of conventions attempted to balance military necessity with humanitarian considerations for combatants. In 1864, the Geneva Convention stipulated that "Wounded and sick combatants, to whatever nation they belong, shall be . . . cared for," and the 1868 St. Petersburg Declaration averred that causing fighters useless suffering should be avoided. The 1899 Hague Convention was the first international codification of war crimes, with twenty-five signatories from Mexico to Siam (now Thailand), including the United States and the major European powers. Prisoners of war were to be treated humanely with no exception for inconvenience. Expanding bullets and projectile gases were banned, as were explosives that could be rained down from above with balloons or other similar innovative technologies. The 1907 Hague Convention addressed the laws of land warfare and attracted more signatories. Some Naval rules were included in the 1907 conventions, covering topics like piracy and blockades. Naval customs had the advantage of their own history, one that was richer and more detailed than for land warfare, independent of the Hague rules which, in 1923, were extended to aerial warfare.

Although some militaries began to build the Hague rules into their training as matters of custom there were no enforcement provisions.

Nor were there provisions in the Hague Conventions for prosecutions of war crimes. The making of war itself was held as a right of sovereign nations, or in other words, politics by other means. Such was the imperial world order, as described in chapter 1. The Augustinian tradition of just war theory, which distinguished *jus ad bello*—justification for making war—from *jus in bello*—conduct during war—was in the background of these diplomatic developments but was not an authority that was cited in the evolving international law.

Despite their limited reach regarding war making itself, the Hague Conventions formed the basis of World War I–era claims about war crimes, but their diverse shortcomings were also revealed. The German zeppelin bombing of London was not a violation because the city had defensive measures, and in any case Germany and France were not signatories. U-boat attacks on enemy shipping were also not a violation. Hague 1907 left ambiguities about what means of delivery of poison gas was banned, and while it prohibited explosives from balloons, Germany did not sign or ratify this declaration, while the US and the UK did. The horrific massacres of Armenians by Turkey did not take place under a declaration of war and were thus also not in violation. As part of the Paris Peace Conference, a postwar commission under the Allies proposed that high-level authorities of Germany and the Central Powers be brought before a tribunal, including the Kaiser. But President Wilson worried about the appearance of "victor's justice," so the US dragged its feet. Wilson died after the signing of the Versailles Treaty.

Momentum for major trials slowed also as various countries assessed their strategic positions in postwar Europe. The impetus for clarifying the many legal uncertainties under prevailing rules weakened as the victorious powers allowed themselves to be satisfied about the outcome of the Great War and were rattled by a global depression. Nonetheless, there were various tribunals concerning individuals who had committed certain wartime acts, though they often resulted in acquittals. In the end, the 1928 Kellogg-Briand Pact that "condemned recourse to war" was perhaps the most important outcome at scale but still fell short of clarifying that the making of aggressive war is itself a crime.[3] Remarkably, it wasn't until the 1949 Geneva Conventions that provision was made for the protection of "wounded, sick and aged persons, children under 15, expectant mothers, and mothers of children under 7" in "hospitals and safety zones," perhaps in recognition of the way modern arms and tactics had created an era of total war.[4]

Getting to Nuremberg

Although heinous acts were committed by Germany during the Great War under the Kaiser, they were at least largely confined to targets with a military rationale, nor were their enemies without taint in this regard. As the Nazis imposed their racist and totalitarian ideology inside Germany in the later 1930s, there was a sense among the Western democracies, at least ostensibly committed to liberal ideals, that the next war threatening to break out would be different. Already in the early years of World War II, reports came of mass arrests, executions, forced labor, denial of food to occupied peoples, and theft of property, and then of large-scale exterminations that singled out Jews, Roma, Slavs, and political prisoners. In 1942, an Allied war-crimes commission signified a process that, later that year, included the League of Nations declaration and a UN War Crimes Commission (UNWCC). Although the Soviet Union did not sign on, that same year the Soviet ambassador to Britain, in a note to the British foreign minister, urged the creation of an international tribunal of major war criminals, and in 1943 Stalin joined Churchill and Roosevelt in the Moscow Declarations that demanded the Axis powers' unconditional surrender and set out some goals for the postwar world, including a United Nations Organization based on national sovereignty and equality. One of the Declarations referenced "evidence of atrocities, massacres and cold-blooded mass executions which are being perpetrated by Hitlerite forces in many of the countries they have overrun and from which they are now being steadily expelled" anticipating judgment "on the spot by peoples they have outraged."[5] Meanwhile the UNWCC had little financial support and hardly any staff, so as the war proceeded it produced scant evidence of crimes, and as an organization it withered away.

To the British, the immense scale of the crimes by the Third Reich's leadership thoroughly outsized the concept of a tribunal that could somehow dignify the Nazi crimes by bringing them under the bewigged pomp of British courts, and in any case the major figures responsible for the war were a matter of fact: Hitler, Goering, Himmler, Goebbels, Hess, and the rest were well known to the world for a decade before the war. Churchill, in particular, favored summary execution, especially in light of the failed attempts at a high-level international trial after World War I. US secretary of the treasury Henry Morgenthau Jr. also argued for summary execution as well as turning postwar Germany into an agricultural society (known as the Morgenthau Plan) so it could never again have the material capacity to wage modern war.

Stalin, however, favored the appearance of a legal process, insofar as he understood it, which was not very far. Though its moral standing might be said to have been compromised by the nonaggression pact with Nazi Germany (and later revelations of events like the Katyn massacre of thousands of Polish officers and intellectuals), it was the Soviet Union that had suffered the most when, as was widely expected, Hitler reneged on the treaty, losing one out of seven of its people. Two members of the Kremlin's inner circle, Vyacheslav Molotov and Andrey Vyshinsky, advocated an international tribunal that would put the facts on the record and establish that the punishments could not be dismissed as victor's justice but valid international law.

(It is no small irony that Vyshinsky prosecuted the Moscow Trials in the late 1930s that purged the Communist Party of Stalin's opponents.)

The Soviet position had been articulated years before by the Moscow University law professor Aron Trainin, an influential Jewish Bolshevik. According to historian Francine Hirsch, "Trainin's 1935 work *Criminal Intervention* . . . argued that the movement to unify international law in the 1920s and early 1930s had failed because the League of Nations had been so preoccupied with halting the spread of communism that it had ignored what Trainin characterized as the real problem: how to prevent 'aggressive war.'"[6] In his next book, *The Defence of Peace and Criminal Law* (1937), Trainin granted the importance of the Kellogg-Briand Pact of 1928 but said it failed to make waging war a punishable offense, urging an international criminal court to try "persons violating peace."[7] Citing Grotius's distinction between wars of defense and wars of aggression, Trainin's concept of "crimes against peace" ultimately proved foundational in the design of the Nuremberg Principles. The UNWCC took the arguments of his book on board, where they reached the White House and President Harry S. Truman himself. But in practical terms, the tribunal the Soviet leadership had in mind resembled nothing so much as the Vyshinsky's show trials, with the sure outcome of death for the defendants. The actual outcome of the major trial, in which several of the accused went free, was not at all what they had anticipated.

Nonetheless, the Roosevelt administration's legal advisors were impressed by Trainin's argument. Somewhat reluctantly, they turned away from the Morgenthau Plan, advocating an Allied declaration that established the criminality of aggressive war as "valid international law." The cabinet secretaries agreed, concluding that the Morgenthau Plan risked turning the Nazi leaders into martyrs. As summarized by Hirsch: "A trial, by contrast, would receive popular support as well as 'the respect of history.' They [the secretaries of state and war and the attorney general] suggested charging the German leaders and their organizations with committing heinous war crimes and with jointly participating in a criminal conspiracy. The latter charge would cover atrocities committed before the war, as well as the crime of planning and waging 'an illegal war of aggression.'"[8] There were, however, many weighty procedural details left to be resolved. The British favored a closed hearing, for example, while the Soviets wanted a public spectacle. The Americans also wanted a public trial, though their goal was to display American values and the rule of law. It was left to the Big Three—Churchill, Roosevelt, and Stalin—during their historic meeting at Yalta, to find some accommodation among their differing goals and conceptions of justice for the major war criminals.

Early in January 1945, the American administration reached an internal agreement that advised Roosevelt that an international tribunal should try "the highest-ranking German leaders" as well as Nazi groups and organizations that implemented the criminal plans. But redrawing the map of Europe was the main agenda item at Yalta, and Roosevelt, near death, put off a commitment to a trial. News of Nazi atrocities, including the murder of American prisoners and the industrialized killing in the concentration camps, increased the pressure. Truman, the new president, charged Supreme Court Associate Justice Robert H. Jackson with preparing "charges of atrocity and war crimes against such of the leaders of the European Axis powers . . . as the United States may agree with any of the United Nations to bring to trial before an international military tribunal."[9]

Before the first trial of the major war criminals could begin, there were still many technical and political problems confronting the IMT. The Anglo-American trial system was different from the French and Russian, for example, to the extent that the Russians didn't know how to translate "cross-examination." Selection of defendants was arduous and resulted in one particularly embarrassing episode in which the wrong person was indicted: The industrialist Gustav Krupp was named in error. He had been disabled since the beginning of the war so it was rather his son, Alfried, who might better have been among the defendants, but it was discovered too late to correct the mistake. (Alfried was tried in one of the subsequent Nuremberg trials, convicted, imprisoned, and pardoned in 1951.) The defendant Rudolf Hess, who had flown a plane to Scotland in 1941in a bizarre attempt to open peace talks but was apparently mentally ill, was an irritant as he had never been tried by the British after years in custody. Some of the issues would have been comical had the matters at hand not been so grave, such as the question of how the judges, one representing each of the Allies, would be garbed in court. It was feared that, if each had worn their customary robes or uniforms, the dais would have looked like a costume party. Finally, all wore black robes except Iona Nikitchenko, the Soviet, who wore his military uniform. It was not the only occasion when he found himself in the minority, which helps explain why the Soviets lost interest in the IMT. Of the nineteen defendants twelve were sentenced to death, though Hermann Goering escaped the hangman by committing suicide in his cell. The three acquittals did not please Nikitchenko, whose delegation was taking orders directly from Moscow. Well before then, the Soviets had lost control of the narrative. The prosecution had wanted the focus to be limited to the Axis crimes, but the Katyn massacre was among the episodes that the defense used to complicate the case: "The language of international law was proving to be a double-edged sword," Hirsch writes.[10] Similar problems of consistency would arise in the case against the doctors.

The other major war crimes trial deserves mention as another instance of US domination of the international scene and of the emergent rules-based order. The main trial in Tokyo, and the subsequent war crimes trials in Japan, are for good reasons far less celebrated than the ones taking place simultaneously in Nuremberg, despite involving nearly a dozen countries, instead of only the four major powers. The Tokyo tribunal operated explicitly under the same principles as the war crimes courts in Germany, though the Americans' insistence that the emperor must not be indicted was controversial. The viceroy of Japan, General Douglas MacArthur, was resolved that Japan would be an ally of the United States in the impending struggle with the Soviet Union; he saw the stability of the throne as critical for that goal. Other significant figures evaded punishment, often going on to important leadership positions in education and medical science. The Imperial Japanese experiments at Unit 731 in Manchuria, which rivaled the Nazi death camps in cruelty, were allowed to pass without prosecution.[11] Thus, the results of the Tokyo trial were far more ambiguous than those in Germany. Modern Japanese political culture has never admitted Imperial Japan's criminality in places like Nanjing, owing partly to the poor foundation set by the trial. As the historian Gary Bass has observed, "Throughout the Tokyo trial the Allies repeatedly asserted that these new principles of international law would apply to themselves as well."[12] But substantive allegations of atrocities by American force members in Vietnam, Cambodia, Iraq, and Afghanistan have gone unexamined.

Allies, Not Partners

I have rehearsed the story about the runup to the first trial at Nuremberg in order to establish that the foundations of the second or medical trial, best known for the Nuremberg Code, were firmly rooted in a difficult but eminently deliberate process among the Allies for the construction of the postwar rules-based order. For the Allies, and especially for the Americans, the two key reference points were the United Nations and the Nuremberg Principles, in the words of Justice Jackson's son, "The twin foundations of an international society ordered by law."[13] Those foundations having been established, and the first trial of major war criminals having implemented them, it would seem that the way was prepared for the IMT to continue prosecuting other Nazi offenders in subsequent trials, including those responsible for medical atrocities, but that was not the way it worked out.

In fact Robert Jackson, the chief counsel, was unenthusiastic about an international tribunal, preferring instead "zonal" trials by each of the Allies for the prosecution of the individuals in their custody. That would avoid the problems associated with the different legal systems that were confronted by the IMT in the first trial. However, the French very much wanted a second trial of industrialists and financiers to be held under the international tribunal: So many of their people had been subjected to slave labor in Krupp factories and others that this would have been an important statement. Jackson disliked the French and Soviet legal structures to begin with, believing that they could send the wrong message to the American people about the legitimacy of the trials. Moreover, he was skeptical about the strength of the case against the Krupps and other major industrialists, fearing that a failed prosecution would tarnish the results of the first case and the whole endeavor. And there was another singularly practical and nonlegal reason for his reluctance: A successful prosecution of industrialists and financiers could be a propaganda victory for the Soviet Union in their struggle against capitalism. The British were similarly unenthusiastic about another trial.[14]

For their part, in the aftermath of the IMT the Soviets were both furious about the acquittal of several defendants and suspicious, as were at least some the French delegates, that the talk of a medical trial was a diversion from the issue of the industrialists who had facilitated the Nazi war machine. And there were real shortcomings, such as the prosecution's decision not to apply the Nuremberg Principles to murders and outrages committed against German citizens. By then the human rights issues had become caught up in emerging Cold War politics. The UN General Assembly contemplated a French proposal for an international criminal court under the Nuremberg Principles, an idea that the Soviets had floated in the late 1930s. In the postwar atmosphere of competition with the US and the British, however, Moscow was no longer so willing to concede sovereignty, in particular as Communist governments were being charged with acts of genocide by former Baltic governments. The Soviets had always viewed the Nuremberg Principles as applying specifically to the Axis crimes, though they gradually suspected them as a device for the West to impose its will on, and isolate, them by means of a system of international law defined by the Americans. These same tensions have rippled through history to the present day, through the Cold War and the collapse of the Soviet Union to Vladimir Putin's fury at the West and visions of a Russian World. Despite the political as well as humanitarian origin of concepts like genocide and crimes against humanity, as the historian Hirsch observes, "None of this undermines the revolutionary nature of the Nuremberg Trials. It does remind us, though, not to idealize a mythical 'Nuremberg Moment.'"[15]

The Path to the "Medical Case"

Jackson's failure to subject Goering to cross-examination was widely noted in the press and by the legal community, contributing to his desire to move on from the four-power arrangement that had been so troublesome. Disappointed, exhausted, and embarrassed, Jackson appointed the ambitious young Army lawyer Telford Taylor as deputy chief counsel in charge of the subsequent Nuremberg trials. Truman was persuaded that the subsequent trials should be solely under American jurisdiction and so informed the British, French, and Soviet governments. The International Military Tribunal was thus succeeded by the Nuremberg Military Tribunals (NMT) under the occupiers' Allied Control Council Law No. 10, providing the legal authority for each of the Allies to try accused persons in their zones of occupation, the basis for the twelve NMT trials in Nuremberg. These were all under US jurisdiction and American judges, applying the legal framework established in the Nuremberg Principles. Although the trial went forward, the circumstances were less than ideal. In the end, as the historian Ulf Schmidt, observes, "The Nuremberg Code probably never received the international recognition which it could have had, had it originated within an international trial."[16]

While debate continued about the wisdom of trials of industrialists and others under various groupings, the American team came to appreciate that a "medical case" made sense as the first trial under the new NMT. Already during the trial of major war criminals, a former concentration camp inmate, a Czech surgeon named Franz Blaha, described medical experiments at Dachau, including hypothermia.[17] The British and the French, as well as the Americans, had both been assembling evidence about atrocities like the Dachau high-altitude and freezing experiments and those at the Ravensbrück camp involving bone transplants and sterilization. The British and the Americans also had access to a number of suspect individuals in their custody. "A trial against German physicians," Schmidt writes, "was also a means to re-educate and denazify German society and explain to the world why and how things had happened."[18]

According to legal scholar Kevin Jon Heller, after much hesitation a key element of the Americans' decision to pursue the medical case was simply that evidence was plentiful. It was also less legally complicated than other options, as it was a matter of fact that these experiments had been done, while the case against the industrialists involved a wide range of crimes from slave labor to murder and to technical issues about whether the law could be applied to corporations and their officers as "juristic persons."[19] By contrast, the medical defendants were assuredly natural persons who either conducted the experiments or did not try to stop them, so if the facts were borne out they were culpable. To organize the evidence, Telford Taylor called on an Austrian-born neurologist who had been serving as a medical advisor in the US Army, Leo Alexander. He proved to be one of the most colorful and influential characters in the Nuremberg tribunals. In 1933, after his father, also a neurologist, was murdered by a former patient with a mental health condition, the young and ambitious Alexander decided to take a "scientific holiday" from his post in

Frankfurt at the Rockefeller-founded Peking Union Medical College. As Alexander was Jewish, he was warned by the family lawyer not to return to Germany because the new regime under Hitler was already enacting its anti-Semitic policies. Though still far from the genocide of the Final Solution, conditions for Jewish professionals were rapidly deteriorating. Through contacts, he secured a position in Boston, later moving to Duke University. Increasingly disturbed by the events in Europe, he tried to join the Army but was rejected as overweight. Finally in 1943 he was admitted based on a waiver and took care of soldiers with "shell shock," what today might be described as post-traumatic stress disorder (PTSD).[20]

As evidence of war crimes emerged, Major Alexander enthusiastically participated in an assignment to help gather more information about medical atrocities. The Nuremberg Principles, complemented by the concept of genocide, constituted much of the legal theory of the medical case. On the evidentiary side much preparatory work for the trial had already been accomplished, though in a somewhat haphazard and coincidental way, by Allied organizations tasked with studying and exploiting German science and technology that had given the Nazis some advantages during the war. Evidence of atrocities in the camps and in factories for rocket production had tumbled out. To help develop the evidence for the medical case and to understand the way Nazi medicine had been subjugated to war aims, Alexander began interviewing a number of senior German military psychiatrists. A neuropsychiatrist by training, the exploitation of his discipline in the murders of people with disabilities was particularly motivating, though Alexander quickly enlarged the scope to the medical atrocities in the concentration camps, of which the evidence was abundant and utterly shocking. Alexander arrived in Munich in May 1945, visiting Dachau and talking with emaciated survivors wandering the streets. American soldiers discovered a fifty-site complex of slave labor camps with Mauthausen at the center, where forty thousand people had died under the most horrendous conditions imaginable.

Alexander's interviews and observations took him through the hellscape of postwar Germany, from Ebensee concentration camp to "euthanasia" sites, to aviation medicine institutes where hypothermia experiments were planned to Dachau where vicious experiments were carried out. As he gathered evidence from his interviews and interrogations, Alexander came to understand the combination of sadism and raw utilitarianism that is now a familiar feature of the degradation of medicine in the Third Reich. In academic psychological circles, personality theory was an important field of study at that time. Harvard's Henry Murray had developed the thematic apperception test (TAT) and used Rorschach, or "inkblot," tests in his assessments, often with Harvard undergraduates he encountered through his role at the university counseling center. Alexander had the idea of performing psychological studies of the major German war criminals, offering his services to the US military and the Rockefeller Foundation, in some ways anticipating Robert Jay Lifton's writings about Nazi doctors. Alexander's personality assessments project gave way to the more immediate issue of the medical war crimes themselves, though it would prove useful in his profiling of the trial defendants.

Alexander produced seven reports for the war crimes investigators on victims of medical crimes, such as those used in the freezing experiments. As planning for the medical case intensified, his reputation was sufficient to gain the attention of Telford Taylor, who had come to appreciate the importance of medical expertise, and especially knowledge of the history of human experiments, for the impending medical trial. Promoted to the rank of lieutenant colonel and given the title "expert consultant to the secretary of war," Alexander was ordered to serve as the advisor to the prosecution of the medical case. Considering Alexander's biography and the expertise of German wartime medicine he had already acquired, rarely has there been a better match between an investigator and their assignment.

Taylor was sensitized to this question by reports of experiments conducted by American and British physicians during the war.

The vice president of the University of Illinois, Andrew Ivy, was appointed as a special consultant to the secretary of war to help the office of chief of counsel for war crimes sort out the problem. Ivy was a highly regarded public spokesman for medical science and a kind of renaissance man: a physician, scientist, musician, and baseball player. Ivy was also willing, in Ulf Schmidt's words, to be "economical with the truth."[21] So, in all these ways he was well-suited to protect the field of medical experimentation from the Nazi taint. In 1946, he attended a Paris conference under French authority with the cumbersome title of the International Scientific Commission for the Investigation of War Crimes of a Medical Nature. Ivy argued that while the commission was developing its own report, the planned Nuremberg trial should continue. In the end, despite the entreaties of other commission members, the United States decided not to participate in future commission activities, preferring to invest its resources in the upcoming trial.

Ivy also set about to develop rules that would distinguish unethical German experiments for ethical ones. Included in an appendix to the Paris report, Ivy listed three major principles in language that would be adopted by the judges months later: consent by volunteers who had been told of the hazards and, if possible, insurance against injury; justification by animal experiments and for the good of society; and avoidance of unnecessary suffering by scientifically qualified persons with no "*a priori* reason to believe that death or disabling injury will occur."[22] Ivy was shrewd enough to appreciate that the question of Allied standards for human experiments would be an important topic for the defense in the trial. The day after the trial began in December 1946, his rather vague guidelines were adopted by the American Medical Association (AMA) House of Delegates and subsequently published in the *Journal of the American Medical Association* (*JAMA*). Months later, testifying as a prosecution witness, Ivy was able to refer to the principles that he had written and whose publication he had ensured.[23]

Leo Alexander, who was also in attendance in Paris, agreed with Ivy that, in the words of a commission press release, "an international, legalized Code of ethics should be published on the use of human beings as experimental subjects."[24] Alexander himself was gradually focusing more on prospective principles of medical ethics, to the extent of writing some diary entries about "voluntary consent," information disclosure avoidance of death and injury, and assessing the risks and benefits of an experiment. These points became an early affidavit submitted to the trial court.[25]

The Trial

Legal scholar Raphael Lemkin, whose book *Axis Rule in Occupied Europe* (1944) explained how Nazi Germany had repeatedly violated the Hague Conventions, had been one of Alexander's colleagues across campus at Duke. He argued that the goal of the extermination of nations and ethnic groups behind the Nazis' European rule was that of changing the "balance of biological forces between it and the captive nations for many years to come."[26] Lemkin called this scheme "genocide," a novel concept underlying the mass murders taking place while he was writing his analysis. The attempt to eliminate an entire people was a crime of such scale that it could not be captured by any traditional categories of criminality. Building on the then widely accepted eugenics framework, this biological strategy implicitly created a special role for medical doctors in "purifying" the Aryan race the Nazis had invented. It also became the principal conceptual thread that tied together the case against the defendants in the doctors' trial.

Another thread was a concept developed by Alexander himself. He called it "thanatology," a term for the murderous practices under medical supervision that implemented genocide. This was critical for the prosecution because although the conduct of experiments on human beings was neither new nor unique to Nazi medicine (a fact that became a significant part of the defense in the trial), thanatology was a step beyond, a "science of killing"[27] that could not be defended by the usual justifications for potentially beneficial human experiments. In a diary entry a few days before the trial started in December 1946, Alexander wrote that "many of these so-called experiments are frankly and openly devoted to methods of destroying or preventing life. . . . This thanatological knowledge supplied the technological methods for genocide, a policy of the German Third Reich, which could not have been carried out without the active participation of medical scientists."[28] Although the concept of genocide is a vigorous legacy of the doctors' trial, thanatology has not been recognized as a standard of international human rights law; the term itself has rather come to be identified with the study of death from various disciplinary standpoints. The genocide theory also limited the prosecution to crimes committed against peoples outside Germany, so crimes like the murders of German children with disabilities were outside the reach of the trial. This, as we saw, was a basis for one of the criticisms of the trials of major war criminals.

Apprehending and identifying the accused was itself no small logistical matter, as there were cases of mistaken identity and others, like Josef Mengele, who escaped or committed suicide. Nor were the chaotic postwar conditions or the attitudes of the defeated Germans particularly helpful. British forces arrested Karl Brandt, Reich Commissioner for Health and Sanitation, who was among the fairly visible members of the last German government in Flensburg. In his official capacity, Brandt was the most senior of the final list of defendants in the trial of *United States of America v. Karl Brandt et al.* He had been implicated in the "euthanasia" program of people with disabilities, chemical warfare, and epidemic jaundice experiments in the concentration camps.

Identification and appointment of the four American judges was difficult in its own way. The positions were far from prestigious.

Interest in Nazi crimes was ebbing among Americans, and for potential jurists leaving their work and homes for many months, both professional and personal sacrifice was required. In the end, the panel was composed of two state supreme court justices from Washington state and Florida, a circuit court judge from Oklahoma, and a former Michigan assistant attorney general.[29]

Finally gathered and held for trial, the indictments of the twenty-three defendants, twenty of whom were physicians, corresponded to the Nuremberg Principles as a conspiracy to commit war crimes and crimes against humanity; war crimes involving responsibility for medical experiments "conducted without the subjects' consent"; crimes against humanity committed "upon German civilians and nationals of other countries"; and guilt for membership in the SS (*Schutzstaffel*), a criminal organization, before the war.[30] Taylor's opening statement described the prosecution case against the defendants for murder, torture, and atrocities in the name of medical science. But, he added, this was not simply a murder trial but reached issues of medical ethics: "Whatever book or treatise on medical ethics we may examine, and whatever expert on forensic medicine we may question, will say that it is a fundamental and inescapable obligation of every physician under any known system of law not to perform a dangerous experiment without the subject's consent."[31]

The Defense

In the event, the prosecution would not unfold as smoothly as Taylor and his team expected. Although it may be all but inconceivable to us that the defendants could see their motives as noble, Nazi medical ethics was part of a worldview that entailed "racial hygiene."[32] The defendants were allowed to choose their own lawyers so long as they hadn't been ardent Nazis. One of the lawyers was Dr. Robert Servatius of Cologne, an experienced and sophisticated legal scholar who, as it happens, later represented Adolf Eichmann in his trial in Jerusalem in 1961. The defense adopted a *tu quoque* tactic to undermine the prosecution, accusing the Allies of the same actions of which the defendants and the Third Reich medical system were accused. They were greatly aided in this approach by the testimony of the Erlangen University psychiatrist and historian of medicine Werner Leibbrand. Called by the prosecution to emphasize that only a few German doctors had been guilty of ethical violations, on cross-examination Leibbrand affirmed that in the case of research involving prisoners, a population he said could never have free will, he would hold any parties to the same ethical standards. If this argument held then ethical violations would in essence collapse any relevant moral differences between Nazi medicine and that of the Allies. Or at least that's how the defense wanted the court to see the matter.[33]

Perhaps the most spectacular case entered into evidence by the defense was a malaria experiment in a federal penitentiary in Stateville, Illinois, involving five hundred male prisoners. Malaria was proving to be an immense problem for the US armed forces in the Pacific, so the hope was that systematic exposure to *plasmodium vivax* would provide information about improved drugs for treatment and to prevent relapses. This was almost certainly the most high-profile human experiment of its kind during the war. Not only did authority for the experiment stem from the White House itself, but while it was in process a team of photographers from LIFE Magazine recorded aspects of the experiment, suggesting the heroism of these incarcerated men who were voluntarily contributing to the war effort in their own way. Featured in the June 4, 1945, issue of the magazine, the experiment was anything but secret, giving the doctors' defense lawyers an easy reference point for their argument, especially in light of a malaria experiment at Dachau. In the final analysis, however, conditions at the death camp were far from those that obtained in the prison; death was not an acceptable end point at Statevillle as it was at Dachau. A few months after the magazine article, Andrew Ivy was able to assert in his testimony that there were ethical rules in the American medical community that would prohibit any such foreseeable outcomes, rules that he himself had drafted and published in *JAMA*.[34]

In the end the judges were not persuaded by the defendants' line of argument, nor that acting under orders from superiors provided legal relief from actions that the individual knew, or should have known, were foreseeably likely to cause injury or death. Formerly a recognized argument in international law, the doctrine of superior orders became a far more limited defense following Nuremberg. A working paper in preparation for the trials specified that "in any trial before an International Military Tribunal the fact that a defendant acted pursuant to order of a superior or government sanction shall not constitute a defense per se, but may be considered either in defense or in mitigation of punishment if the tribunal determines that justice so requires."[35] Whether there is any place for the doctrine at all in international criminal law is still not fully resolved, but what the Nuremberg tradition makes clear is that genocide or crimes against humanity cannot be justified by superior orders. This shift is an example of the consolidation of centuries of evolving legal doctrine within the ambit of the postwar order. In any case, the allegations faced by the defendants were so gross, their defense so preposterous, and the evidence so vast and public that any superior orders defense would be risible. "At Nuremberg itself," writes a scholar in an ICRC paper, "the crimes alleged were of such magnitude that the absolute nature of the denial of the superior order defence [sic] made little or no difference."[36] Brandt and six others were sentenced to death, nine more were sent to prison. It was well understood that no punishment could be congruent with the crimes, that in such a case justice could not be complete, but neither could the world afford no response at all.

Although, as we have seen, there was not universal enthusiasm among the Allies for a medical war crimes trial in Nuremberg, there was evidently sufficient momentum and interest triggered by the first trial such that ultimately the case was pursued. The International

Military Tribunal for the Far East (IMTFE) was ostensibly modeled on the IMT in Germany. But Japan was occupied by a single power, the United States, and China, where the horrific experiments conducted by Unit 731 had taken place, did not file a complaint. In addition, Harbin, where Shirō Ishii's complex was located, was remote. With indictments based on the Nuremberg Principles, the prosecution of twenty-eight of Imperial Japan's civilian and military officials began in 1946, but the final ruling was not issued until 1948. That would have given the judges months to consider the relevance of the Code for Japanese medical war crimes, if they had known about them. Despite awareness of the Nuremberg prosecutions then being carried out, in 1947 the minutes of a group of US military representatives concluded that "the value to the U.S. of Japanese BW [biological warfare] data is of such importance to national security as to far outweigh the value accruing from 'war crimes' prosecution."[37] American intelligence services had obtained information about Soviet biological weapons programs during the war, and there was no better source of data about human effects than that collected by the Japanese. When a 1981 article in the *Bulletin of the Atomic Scientists* finally cast light on the story, Judge Bert Röling from the Netherlands said that "as one of the judges in the International Military Tribunal, it is a bitter experience for me to be informed now that centrally ordered Japanese war criminality of the most disgusting kind was kept secret from the Court by the U.S. government."[38]

Before the Code

The case for the rules-based order as a determined effort to create an explicit framework for international relations is exemplified by the choices made by the prosecution, and ultimately the Nuremberg judges themselves, within the domain of the conduct of human experiments. Their decision to render the ten-part statement indicates their dissatisfaction with the prewar status of rules for human experiments that the new rules-based order required a more explicit statement that would be internationally recognized. Yet there were precedents from which to draw. Experts like Andrew Ivy and Leo Alexander, as well as noted German legal scholars like the defense attorney Robert Servatius, would surely have been aware of at least some of the historic cases and conventions concerning human experiments, both in the United States and in Europe. Most prominent of these episodes was that of Walter Reed, the US Army physician assigned by the Army surgeon general to take a team of doctors and nurses to Havana in 1900 where a yellow fever epidemic was raging. Through a variety of ingenious, challenging, risky, and sometimes literally nauseating studies, the Reed group identified the yellow fever vector as a silver-backed female mosquito, *Aedes aegypti*. Besides this dramatic moment in the history of public health, a project in which a physician and a nurse died of the disease, Reed is also credited with requiring at least some of the experiment volunteers to sign a contract that set out the risks and benefits of participation, albeit one that hedged on risks, invited the naïve reader to infer greater benefit than could reasonably be expected, and made a substantial offer of gold to those who agreed to take the bite. (These incentives are still matters of some struggle within research ethics committees, though the contract did have the virtue of being confined to a single page.).[39]

Also in 1900, following a scandal involving the inoculation of serum from syphilis patients into women and adolescent girls in hopes of finding a preventive measure, and under pressure from an incipient mass news media and progressive social democrats, the Prussian Ministry for Religious, Educational and Medical Affairs issued a directive that placed responsibility on the relevant medical director of an institution:

> That medical interventions for purposes other than diagnosis, therapy, and immunization are absolutely prohibited, even if

> all other legal and ethical requirements for performing such interventions are fulfilled if: (1) the person in question is a minor or is not fully competent on other grounds; (2) the person concerned has not declared unequivocally that he consents to the intervention; (3) the declaration has not been made on the basis of a proper explanation of the adverse consequences that might result from the intervention.[40]

The powerful Prussian bar embraced the directive, but the medical profession was far less enthusiastic. The directive had little impact in Prussia or in other German states. The pertinence of such rules in Germany intensified due to the advanced state of medical science and human studies as compared to Britain and France. A second attempt to control human experiments unfolded in the 1920s when the physician Julius Moses, a Social Democrat member of the Reichstag, reported that seventy-five children had died in Lübeck following experiments with tuberculosis vaccinations. In 1931, the Ministry of the Interior published "guidelines on innovative therapy and scientific experimentation involving human subjects." Among the fourteen points were the priority of animal testing, documented consent, special rules for children and the dying, and a requirement that medical ethics be taught in medical schools.[41]

Although these guidelines are in some ways more advanced than the Nuremberg Code and the Helsinki Declaration, they had an uncertain legal status. Were they binding law throughout Germany, merely an administrative action, or part of the legal contract with Reich physicians? Despite their impressive detail and elements that anticipated and even exceeded postwar standards, this ambiguity may explain why the guidelines were not a basis for the Nuremberg prosecution, nor were they cited in the 1948 Omnibus Law passed by the Reichstag that brought hundreds of laws and regulations from the Reich into the new Federal Republic of Germany.[42] At the practical level the guidelines were not widely respected even at the time. Contextually, the paternalistic German medical tradition, emerging enthusiasm for eugenic social medicine, and the economic and political crisis of Weimar Germany did not provide fertile ground for the guidelines. Their principal author and advocate, Julius Moses, died in the Theresienstadt concentration camp in 1941.[43]

As other writers have argued, the prosecution intended the doctors' trial to be an explicit rejection of Nazi medical ethics.[44] This was clearly the prosecutors' immediate goal. But as I have argued, the larger context embedded medical science under the ambit of a new rules-based international order, focusing in particular on the highly symbolic practices of human experiments. For all its fits and starts the trial succeeded in reaching that goal, though the influence of the Code itself is too easily couched in anachronism. Historians generally agree that it had little impact in the United States or in Britain. The psychiatrist and Yale law professor Jay Katz says the general reaction in American medicine at the time was that "it was a good code for barbarians but an unnecessary code for ordinary physician-scientists."[45] Similarly, in her review of the Code's impact in Britain, historian Jenny Hazelgrove concludes that "in 1946 the medical elite had successfully negotiated the threat posed to the status of medical science and the progress of research by public abhorrence of Nazi science."[46] It wouldn't be for the better part of a generation that the Code would assume its now familiar iconic status, and that would come in light of later events amid vast social change in the 1960s and 1970s. In the meantime, what Hazelgrove calls the medical elite were by no means ready to cede their authority over the conduct of human experiments under the new rules-based order.

3
The American Stamp

Long before Nuremberg there had been periodic debates about the limits of physician discretion in relation to experiment participants and patients more generally. Jay Katz observes that the history of medical ethics, from Hippocrates through the middle of the twentieth century, was one that concentrated on physicians' duties to patients according to the "harm principle."[1] Nineteenth-century French physician Claude Bernard asserted that no experiment can be justified if it might harm the subject.[2] This deontological approach continues to be represented in modern bioethics in principles of nonmaleficence and beneficence. Medical deontology was the term used to describe medical ethics in much of prewar Europe and remains so today. But as an ethic of the physician's duties, deontology did not necessarily include attention to the patient's wishes, an omission that sustained what has come to be described as medical paternalism. This physician-centered tradition of medical deontology was disrupted by the cataclysmic moral failure of German medicine: A distinguished medical culture in an advanced European country took an active role in legitimizing medicalized murder on a massive scale. In stretching physician discretion far beyond the limits of justification and humanity, the Holocaust opened up an unprecedented space for a consent standard, despite the fact that consent is far from the main point regarding the crimes of Nazi medicine.

Harm or Consent?

Not that consent to a medical procedure arose suddenly or with no precedent after World War II. A consent standard appeared in English law as early as 1767, though it had more to do with warning a patient of a potential bad outcome than permission to proceed.[3] In Prussia, following the Neisser case, the noted neurologist and psychiatrist Albert Moll published a list of six hundred experiments he considered ethically questionable in his book *Arzlich Ethik* (1902). Moll argued that the doctor-patient relationship should be seen as a contract, with the patient fully informed and their self-determination respected. His views were cited by lawyers but not were not so well-received by his colleagues.[4] In 1917, the Polish physician Teodor Heiman harshly criticized his colleagues for reducing their patients to experimental material, exploiting their advantage in the power dynamic with people who are dependent on them and fear being abandoned.[5] The doctors' trial made evident the degradation of German medicine, a culture that had been widely viewed as the pinnacle of the profession. Providing posterity with that sensibility about medical ethics as part of a new international rules-based order was an important goal of the Nuremberg judges.

The British physician and World War II veteran John Pridham reflected on "the failure of German doctors to attempt to combat the Nazi ideology which led to both professional and national debasement, and their own profession becoming disintegrated."[6] With the leadership of Pridham and others, in 1948 the new World Medical Association (WMA) endorsed a Declaration of Geneva that was recommended as an oath to be taken by all physicians. Remarkably, the new West German medical association only "grudgingly accepted the role of penitent scripted for them by the WMA," insisting that Nazi medical crimes applied to a small minority of German physicians.[7] However, the struggle for professional control over the exact contours of the ethics of human experiments, both the details and the controlling authorities, was only beginning.

The centuries-long contest between the standards of harm and consent concentrated mainly on experiments rather than regular medical care more generally. Deontologically oriented commentators did sometimes note the importance of consent in what is today called the therapeutic alliance with the patient, especially when drastic surgeries like amputation were involved or in practical terms when family members' preferences might be considered at the bedside when a doctor attended in the patient's home. Nonetheless, years after the changes associated with the doctors' trial, young medical students learned that occasionally the truth should be withheld from patients, justified by a doctrine called the therapeutic exception, also known as the therapeutic privilege. Like the routine application of a widely recognized therapy, experimental procedures judged to be within the best interests of the patient, or at least not in conflict with them, were generally regarded as within the bounds of medical discretion.

But the notion that duties to patients should specifically include respecting the informed patients' wishes, a thoroughgoing consent standard that distinguishes modern bioethics from traditional medical ethics, was not embraced for decades after Nuremberg. Or if it was, it was treated as a "formal" requirement that was often observed in the breach. What actually happened in the confines of the one-on-one consulting room for all those centuries we cannot know; indeed, it seems likely that the very notion of truth-telling is far from universally applied in Western medicine, let alone in cultures in which it is actively rejected. What we do know is that the formal study of medical deontology, with its emphasis on the doctor's duties rather than the patient's rights, did not suddenly change after Nuremberg. Especially interesting in this regard are some pieces of evidence from

Eastern European countries whose new Communist governments repressed contact with the West. Despite that isolation, these were after all the countries that were scenes of some of the worst atrocities and devastation, countries caught between the armies of Nazi Germany and the Soviet Union, often referred to by historians of the period as "bloodlands." The doctors who saw the horrific injuries to both combatants and civilians wrought by the war in their own backyards, and those were themselves imprisoned in the camps, cannot but have been deeply affected. Within weeks of the end of the war, in June 1945, a professor at Charles University in Prague, Bohuslav Bouček sent a lengthy handwritten letter to his dean proposing the creation of an Institute of Medical Deontology. According to the official history of the medical humanities institute in Prague, he "was prompted to this activity by his personal experience with the misuse of medicine in the concentration camps where he was detained for almost the whole war." Before the war Bouček was a well-known advocate of reforms in Czech medical education. Among other subjects, with the advent of Bouček's institute, medical students were to be taught about "eugenics" and "euthanasia." The Institute appears to have ended with his death in 1953.[8]

The deontological tradition in this region appears to have continued its hold in medical education throughout the Communist period, exemplified by the Polish Deontological Code of ethicist Tadeusz Kielanowski in March 1959. Although Kielanowski's code included special protections for vulnerable research participants, it is noteworthy that operating within a socialist system it also made special reference to the needs of society, unlike the Nuremberg Code with its emphasis on the absolutely essential right to voluntary consent.[9] Nor is it surprising that the historic deontological tradition had such staying power in these societies through the Cold War, while more modern patient-centered interpretations of medical ethics were gradually emerging. In article 6 of the Hungarian Deontology Code from 1959, for instance, physicians were obliged to obtain the written consent of patients before performing a medical intervention such as an operation, except in a medical emergency when interventions could be carried out without the patient's expressed consent.[10] The 1955 French Code of Medical Deontology likewise asserted that doctors had a duty to consult with patients and respect their "wishes" and "objections," even if these ran counter to their advice. Suspending this rule was only permissible in extreme emergencies or if the patient was incapable of giving consent, thus highlighting the power patients had over the health and integrity of their body.[11] This is despite the fact that the French medical ethics regime was quite paternalistic, such that doctors could withdraw treatment from a noncompliant patient if they so wished.[12] Moreover, George Burguignon's Code of Ethics for Research on Humans, also from 1955, an updated version of Claude Bernard's nineteenth-century medical ethics principles formulated in response to Nazi crimes, stressed special protections for captive populations. The decision to conduct human experiments ultimately rested with the "conscience of the experimenter," irrespective of how important it seemed for science and society. Here, as in Kielanowski's code, participants had to be fully "informed" and "free to accept or refuse" their involvement in clinical trials, thus highlighting the importance of voluntary decision-making and the inviolability of bodily integrity.[13]

Reasserting Medical Authority

In the face of official restrictions, the medical and scientific communities in the Eastern countries struggled to maintain contact with Western counterparts. The organized medical profession in the former Communist states did not become functioning members of the WMA again until after the fall of the Berlin Wall in 1989. Meanwhile, in the West one indication that the winds of medical ethics were shifting with the advent of the new rules-based order is the way that organized medicine responded to changes in attitudes toward human experiments, the most vivid example of physicians' practices with people in their care. There was no single source for the new ethical pressures on the international medical establishment. Partly, the impetus came from advances in medical science itself that made human experiments both more useful for the growth of knowledge and greater rigor in the conduct of those experiments, with the controlled clinical trial an emerging gold standard. Partly, it was a result of reactions to the conclusions of the doctors' trial and the subsequent creation of both hard and soft international law. The Preamble to the Universal Declaration of Human Rights reminded the world that it originated as a reaction to "barbarous acts which have outraged the conscience of mankind."[14]

Just six years after Nuremberg, why would the leaders of the international medical community see a need for another set of standards to govern human experiments? The very fact that the Code was authored by judges rather than physicians unsettled the medical leadership who wanted to assert their own professional self-regulation. This, despite the fact that the Code's core ideas seem to have been suggested by the physician experts Andrew Ivy and Leo Alexander. "By the early 1950s," Schmidt writes in an unpublished article, "the WMA became increasingly irritated by the public declarations of mostly francophone medical lawyers who campaigned for the creation of binding international medical law in the form of a *Code de droit medical international*."[15]

The WMA adopted an International Code of Medical Ethics and in 1950 adopted a resolution condemning euthanasia. But the pressure on the WMA to go further continued to build. Belgian and Swiss commentators threatened to undermine the WMA's authority with a study of international medical law by the WHO and authorizing governments to regulate physicians' conduct in times of peace and war. Rejecting the views of organizations like the Medico-Juridical Commission of Monaco and the International Committee of Military

Medicine and Pharmacy, "the WMA maintained that doctors, in juxtaposition to governments, were guided by their 'conscience' in practicing medicine, while 'government is, by definition, without conscience.'"[16] In a direct response to the Monaco Commission, in 1952 the WMA warned that any document drafted by the Commission would be rejected by the international medical profession. But the pressure on the WMA continued to build as dozens of other organizations, such as the Red Cross, the International Council of Nurses, the French Academy of Medicine, the Societe de Medicine in Paris, and the Dutch Royal Medical Association insisted on the need for a code of international principles of medical ethics. The more the WMA asserted its special authority and expertise on the ethics of human experiments, the more pressure was building for it to step up.[17]

Finally in 1954 the WMA began what turned out to be the lengthy and arduous process of drafting an ethics code for human experiments, one that immediately ran headlong into internal controversy. A draft offered by the WMA's Committee on Medical Ethics recommended that healthy participants be fully informed about an experiment, which would have included disclosure about whether they were in a placebo arm. After World War II placebo controls were an important part of more rigorous methodology for the conduct of clinical trials. This proved to be a sticking point and not only, as Susan Lederer has explained, for the influential American member.[18] The disclosure to healthy human participants of their assignment to a placebo arm was undesirable to the Danish member of the medical ethics committee as well. Why such a disclosure would be a problem according to the thinking of these members was not clearly articulated by them at the time but was surely owing to the way that fully informing participants of the fact that they were in the placebo group could undermine the rigor of the experiment. Others, including the editor of the *British Medical Journal,* shared this reservation about fully informing the healthy members of a placebo control group. Notably there was no equivalent discussion with regard to sick patients, presumably because the precedent of concern was the Nazi concentration camp experiments and experiments in regular prisons, which would normally have involved healthy people.

At the same time, the WMA felt its territory threatened by ICRC and WHO efforts to define wartime standards for doctors and medical scientists. In the face of sentiments from UN organizations that it felt were openly "hostile" toward the medical profession, the WMA finally aligned with the ICRC, the WHO, and the International Committee on Military Medicine and Pharmacy (ICMMP) to adopt a new code of wartime medical ethics in 1956.[19] A further problem was the definition of a medical experiment: Are experiments not intended for patient benefit governed by medical ethics or by law? At first the WMA advisors hedged, but ultimately determined that the organization could not both leave the ethics of human experiments to the lawyers and yet claim professional jurisdiction over medical conduct. As Schmidt concludes, "The only viable solution left was the decision, probably taken reluctantly, to draft a new code (or codes) of ethics to adequately address the subject, and thus allow the WMA to regain the initiative in the debate."[20] That decision was made official in 1960 with two reports from the medical ethics committee, clearing the way for a draft declaration. Ultimately, the WMA was both informed by the controversies of the 1950s, able to hold its ground as the authoritative body in the field of human experiments, and able to shift the domain in favor of medical discretion. But what remained for a consent standard? And how far would the WMA's code differ from Nuremberg?

The 1964 Declaration

Insistence on professional self-governance, especially in opposition to legal or regulatory oversight, is a persistent theme in the history of human research ethics. The WMA's discomfort with the substance of the Nuremberg Code and its dissatisfaction with its juridical rather than medical authority was also reflected in the deliberations of the association's Committee on Medical Ethics. Working toward what finally became the first draft of the Declaration, in 1960 the committee opposed "hard and fast" rules to govern human experiments (a common sentiment among many leading physicians including some of the profession's critics like Henry Beecher) but nonetheless proposed certain restrictions. On the whole, people in dependent relationships with investigators (e.g., medical students) and persons held captive (e.g., persons in prison) should not be recruited, according to the WMA committee. Later on, prisoners of war were added to the list of those who were off limits but not without dissent among the members. Even among physician leaders, there was often disagreement about certain populations. Conflict among committee members also ensued over the exclusion of children and about the impression that the wording of a draft code of ethics could make on the public.

As to persons incarcerated or institutionalized, the controversy over "captive subjects" was intensified in 1962 by British physician Maurice H. Pappworth who published an article called "Human Guinea Pigs: A Warning" in *Twentieth Century* magazine with seventy-eight examples, followed by his influential book *Human Guinea Pigs: Experimentation on Man* in 1967. Nonetheless, defenders of prison research emphasized their benefits as aiding in prisoner rehabilitation. Finally, the draft presented to the WMA council by the medical ethics committee in 1964 called "Recommendations Guiding Doctors in Clinical Research" did not mention institutionalized children or prisoners. The Declaration ultimately passed by the WMA General Assembly meeting in Helsinki in 1964 required written consent from an informed healthy subject. However, in the case of patients who were receiving clinical care as well as being involved in an experiment only verbal consent was required "consistent with patient psychology," the latter provision a gesture toward the "therapeutic exception," the view that physicians may need to withhold bad news from patients under their care. Also called the "therapeutic privilege," modern bioethics frowns on physician discretion in information disclosure, which is certainly not to deny that it often happens in practice.

Some of the Nuremberg Code statements are preserved in the 1964 Declaration, such as the volunteer's right to withdraw from the experiment at any time. However, several elements that appear in the Code are missing in the 1964 Declaration, amounting to substantive and critical differences between the Code and the Declaration. One missing element in the Declaration is the Nuremberg requirement that "the voluntary consent of the human subject is absolutely essential," modified in the Declaration so that consent may be given by the "legal guardian" in cases of "legal incapacity." The absence of such a mechanism in the Code would obviously have ruled out many potential research populations that could benefit from studies, thus limiting experimental discretion. A second abandoned marker is the Code's statement that "during the course of the experiment *the human subject* should be at liberty to bring the experiment to an end if he has reached the physical or mental state where continuation of the experiment seems to him to be impossible" (emphasis added). Instead, the Declaration provides that "the investigator or the investigating team should discontinue the research if *in his or their judgement* it may, if continued, be harmful to the individual" (emphasis added). Here professional *discretion* is established in a forceful fashion, unlike in the Code which is emphatic about professional *responsibility*.[21] The third Code requirement for prior animal experimentation is also missing from the Declaration as it was considered "too restrictive."[22] Just before the final language was adopted in June 1964, a draft clause that restricted experiments with individuals in a "dependent relationship" to the experimenter was watered down. Also at the last minute, the title of the Declaration was changed from "Ethical Principles Guiding Research Workers in Clinical Medicine" to

"Recommendations Guiding Doctors in Clinical Research," thus freeing those without a medical degree from complying with human research protections. With this seemingly modest change, while the Declaration was an assertion of the medical profession's self-policing it also freed the WMA from legal arguments that it was exceeding its scope of authority beyond medical doctors.[23]

Susan Lederer suggests that drug development interests exercised significant leverage from the very origins of the WMA. American financial support was crucial to the organization's creation and survival in the early years, including considerable contributions from the pharmaceutical industry, which included many of its leaders as members of the WMA's US Committee, also maintaining WMA headquarters in New York City. Concluding that the early Declaration bore a firmly "American stamp," Lederer writes that "the AMA delegates to the World Medical Association . . . who worked closely with pharmaceutical company executives, recognized a potential threat to American drug development by restrictions on the use of prison inmates."[24] Though it is now highly restricted, US prison research thrived until the early 1970s and was a favored site of pharmaceutical industry studies.[25] On this account two imposing forces shaped the lengthy process that eventuated in the first draft of the Declaration: the medical establishment and American interests. Some saw the hand of the United States in certain of the less restrictive elements of the Declaration. Even America's friends drew unflattering conclusions about its delegates' "insidious influence" on the final draft. In the Declaration's seemingly permissive attitude toward prison research, an editorial in the *British Medical Journal* by a writer using the nom de plume, Pertinax, also noted a pullback from the stricter spirit of the Nuremberg Code, despite the fact that the Code had been promulgated in large part by the United States: "One of the nicest of the American medical scientists I know was heard to say, 'Criminals in our penitentiaries are fine experimental material—and much cheaper than chimpanzees.' I hope the chimpanzees do not come to hear of this."[26]

Separating from the Declaration

After the collapse of the Soviet Union, US hegemony in the rules-based order spiked for a decade and then began a gradual decline as other players became more prosperous, and the US made a series of catastrophic strategic errors, like the war and subsequent occupation of Iraq. In the first years of the twenty-first century it turned out to be exceedingly difficult for the US to sustain a unified, dominant authority as more global actors entered the picture. That was also the case in medical science when global health disparities were a matter of growing concern. Thus, liberal principles implicit in the international order were put to the test. A case study of cracks in that hegemony can be perceived in the early twenty-first-century debate about revisions to the Declaration.

If the medical-industrial complex was in fact the sleeping giant underlying the American stamp on the Declaration, that giant was awoken by the 6th edition adopted amid unprecedented controversy in Edinburgh in 2000. Prompted by ethical challenges encountered in HIV/AIDS trials, there were two points that drove the controversy: what standard of care was required in developing countries in clinical trials with persons who are ill (paragraph 29), and whether continued access to "the best proven prophylactic, diagnostic, and therapeutic methods identified by the study" should be guaranteed to the participants (paragraph 30). Concerning paragraph 29, should the worldwide standard of care prevail even though it may not be available in the host country? Should it prevail even when the illness to be addressed is trivial and a more effective treatment is being sought? A third issue, what obligations a research entity had to participants and to the community after the study concludes, was also referenced in the ensuing controversy.

Restrictions on placebo controls appeared in the 1996 version of the Declaration. In 2001, the FDA signaled its reservations about the Declaration's new limits on the use of placebo controls by referring only to the 1989 version (FDA 2001). In 2000, two FDA officials explained the concern. Writing in the *Annals of Internal Medicine*, Robert Temple and Susan Ellenberg pointed out that the phrase "best proven" would ban "not only placebo-controlled trials but also active-control and historically controlled trials." Rather, they argued, "for conditions in which forgoing therapy imposes no important risk . . . the participation of patients in placebo-controlled trials seems appropriate and ethical, so long as there is informed consent."[27] R. K. Lie and three other bioethicists had a somewhat different complaint in an article in the *Journal of Medical Ethics* in 2004. They noted a modification to the 2000 version that "apparently allows placebo control trials in any situation where there is a 'compelling and scientifically sound methodological reason' for their use." But Lie and his coauthors argued, not even the strongest advocates of placebo controls have claimed that this could be the only condition; surely at least some other conditions should come into play, they contended, like not subjecting participants to serious or irreversible harm. The WMA's failure to correct for this obvious error in the several years since the 2000 draft led Lie and the other three bioethicists to the conclusion that the Declaration had lost its "moral authority." They then surveyed five other research ethics policy documents from national and international groups. Unlike Helsinki, "all affirm that, under certain conditions, it is ethically justifiable to conduct a trial in a developing country in which the participants are provided medical interventions that are less than the worldwide best standard of care."[28]

Opposition to paragraph 30 was founded on the inadequacies of many health care systems that would in effect tie the hands of drug companies and require long-term programs of health care that they found unaffordable. Writing in the WMA's journal, a Dutch critic argued that "The dispute about paragraph 30 showed a gap between North and South, between wealthy nations and poor nations," as to what the South had a right to expect, and the North had a duty to provide. To him and others who supported the revision, it was a straightforward moral question: "The time has come for both the WMA and for the pharmaceutical industry to decide on driving either on the left side or on the right, in other words, to choose in favor of the rich or the poor."[29]

Again, the not-so-hidden hand of the US drug industry was presupposed as the main obstacle to maintaining the universality of the Declaration of Helsinki. A strongly worded defense of the revised paragraph 30 appeared as an editorial in the *Canadian Medical Association Journal* in 2004: "In wealthy countries, continuing access to study interventions are generally assured by the near-universal availability of health care. This is rarely the case in the developing world; thus, the need for a clear statement that places responsibility for providing continuing treatment of study subjects on researchers and their sponsor." Leaving no doubt as to the source of the trouble, the editorial continued that objections to this principle were raised "notably by the U.S. Department of Health and Human Services and multinational pharmaceutical companies. . . . [The proposed changes] allow researchers and their sponsors to weasel out of providing continuing care to study subjects wherever administrative, economic or political circumstances create difficulties."[30]

"First Among Equals"?

In 2004, the FDA proposed that clinical trials conducted outside the US that are not conducted under an investigational new drug (IND) application need no longer comply with the Declaration. As the world's most powerful sponsor of drug research appeared to be signaling its strong dissatisfaction with the Declaration, during 2005 and 2006 efforts were made to rescue its singular status. By then, with the FDA recognizing only the1989 version, there seemed little room for compromise. European bioethicist and journal editor Udo Schuklenk said that "[the US FDA] never liked the Declaration of Helsinki and worked overtime to get it changed. When they failed, they simply decided to ignore it."[31] Though agreeing on the ultimate conclusion, lawyer bioethicist George Annas's analysis of the historical relationship between the FDA and the Declaration was somewhat different. In his view, the Declaration was favored as long as it remained weak and was not brought under scrutiny during the AIDS era: "For the last 30 years [US interests] said they loved the Declaration of Helsinki because it's not as strict as the Nuremberg Code, which says you can't experiment without consent, end of story." Annas accused the FDA of walking away from the Declaration when advocates lobbied to change it after the AIDS trials: "[It's] just totally hypocritical on their part to follow the DoH [Declaration of Helsinki] as long as it says what they want it to say, and as soon as its changed, say it doesn't mean anything."[32] Another possibility is that the agency never paid much attention to the Declaration in the first place because it seemed so benign that objecting to it was akin to rejecting motherhood and apple pie. Until, that is, the issues raised by the HIV/AIDS trials.

For their part, FDA officials denied the motives attributed to them but thought that the proposed reforms in the Declaration were the wrong way to implement change. Temple said that, like the WMA, he was upset that people in poor countries don't get good medical care: "But I don't think that undermines the ethics of a trial." The Declaration, Temple said, "moved from a purely ethical document to a document that is increasingly interested in social justice."[33] Although this statement was perhaps not the most felicitous defense of the US position, it does illustrate a basic difference in world views, one that emerged again during the COVID-19 pandemic: Should historically underserved communities be given priority access to ventilators and other interventions that may need to be rationed? In the US it is common for ethics standards to be distinguished from premises that set the conditions for social reform; amid the COVID-19 pandemic, a group of American bioethics program directors agreed the health care system "cannot remedy the structural inequalities of the social system in the U.S."[34] For many others, especially for those in the developing world, these goals cannot be differentiated. In response to the program directors, Jonathan Marks, citing other options, argued that social justice in matters like triage should not be so easily set aside: "Every decision our institutions make (or fail to make) in planning for and responding to a pandemic is a decision about the kind of society we wish to build in the wake of that pandemic."[35]

Perhaps anticipating that the US would still not be satisfied and concerned with the Declaration's lasting influence, the 2008 drafters seem to have decided to dig into their position that the Declaration was, in the words of two commentators, the "first among equals" in research ethics and regulation, asserting in the Introduction that "no national or international ethical, legal or regulatory requirement should reduce or eliminate any of the protections for research subjects set forth in this Declaration."[36] In its 2008 version of the Declaration, the WMA made some of the modifications that the opponents insisted upon. Nonetheless, that same year the FDA published its final rule on "human subject protection; foreign clinical studies not conducted under an investigational new drug application."[37] As was widely anticipated, instead of the Declaration, the rule specified "that such studies [i.e., those not conducted in the United States] be conducted in accordance with good clinical practice (GCP), including review and approval by an independent ethics committee (IEC) and informed consent from subjects." The GCP standards are derived from The International Conference on Harmonization of Technical Requirements for Registration of Pharmaceuticals for Human Use (ICH), a conference of the European Union, Japan, and the United States. In defending the switch from the Declaration, the FDA said that it intended to avoid redundancy in regulatory requirements and subsequent needless delays in the approval of new medications. The agency also cited the greater detail provided by GCP and the confusion caused by the periodic revision of the Declaration.

In spite of these justifications, some critics tied the FDA's jettisoning of the Declaration to the ideology of American exceptionalism in foreign affairs expressed by US President George W. Bush and allegedly carried through in resistance to international agreements on climate change, biological weapons, nuclear tests, and land mines, among other issues. In other words, they viewed the FDA's action as an exercise of American hegemony in a supposedly rules-based order, an exercise of power with precedents in other international domains. Bioethicist Stuart Rennie concluded that "the FDA decided to adopt the GCP during the Bush administration, in a climate marked by reluctance to engage with any international regulation not suiting U.S. interests," and urged the incoming Obama administration that it "should adopt research ethics regulation that places the interests of participants and communities, as well as concerns about social justice, back in the center of the picture, where they belong."[38]

Consensus and Conspiracy

There are at least two important differences between the authority of the Declaration and that of the GCP. The first is that unlike the Declaration, the GCP had to be approved by the regulatory authorities of all the participating countries. However, the GCP is only a guidance document which allows for flexibility in the conduct of clinical research, and a guidance cannot be the basis of regulation. Nonetheless, noting that the GCP does not provide substantive guidance on the use of placebos or on the provision of effective drugs to study participants once a trial is over, writing in the *Lancet,* Peter Lurie and Dirceu Greco saw in the policy changes not only neoconservative government ideology—the suggestion that US foreign policy exceptionalism created an environment for American arrogance in exploitive drug testing—but once again the hand of the pharmaceutical industry. In a direct response to the 2003 article by Lie and coauthors in the *Journal of Medical Ethics*, Lurie and Greco suggested:

> The Department of Health and Human Services and its daughter agencies, ably assisted by the U.S. drug industry, have crafted a strategy that has reached its apotheosis in the current FDA proposal. In a particularly Orwellian touch, the National Institutes of Health . . . recently relied heavily on ethics documents developed in the U.S. and the U.K., as well as the heavily U.S.-influenced document from the Council for International Organizations of Medical Sciences [apparently a reference to GCP], to declare that the Declaration of Helsinki does not represent the "consensus view" on the use of placebos in developing-country trials.[39]

These allegations recalled the earlier history of US and pharmaceutical industry influence over the Declaration and the WMA described by Lederer. How well did they stand up? To their claim that the US government had conspired with the drug industry in citing non-Declaration documents as the true "consensus view," Lurie and Greco cited only the 2003 paper by Lie and coauthors. But the authors of that paper were a Norwegian academic and the other three, though National Institutes of Health (NIH) employees, were not writing as government representatives nor were any employed by pharma. In itself, that list of authors does not provide strong evidence of a conspiracy. Of the five documents cited by them, which Lurie and Greco contended were US- and/or UK-based or influenced, one was from the European Union's European Group on Ethics in Science and New Technologies.

Of course, conspiracies of interest don't necessarily require literal consultation in smoke-filled rooms. Lurie and Greco were on firmer ground when they disputed the notion that these documents represented the true "consensus view," especially since Lie and coauthors allowed that "there is presently no worldwide consensus opinion on this [the placebo control] issue," which is seemingly at odds with the title of their paper on "the International Consensus Opinion."[40] And, as Lurie and Greco pointed out, the Declaration must be approved by an assembly of eighty-two medical organizations, whereas the GCP guidelines are approved by just six parties that do not include the developing world. Just as Laurie and Greco's invocation of George Orwell's nightmarish vision of a conspiratorially reconstructed past seems to have been a bit of rhetorical overreach, much the same must be said of Lie and coauthors' insistence on a consensus that they themselves depreciate. If nothing else, the tough talk on both sides illustrates that the strong emotions aroused in this dispute sometimes generated more heat than light.

Hard Bioethics

In spite of the hopes of some advocates for a new approach during the Obama administration, the FDA position remained unchanged. In a 2014 interview, FDA official Temple reiterated the view that placebo controls can be ethical and expressed approval of the 2008 revision. References to the Declaration in the FDA rules on foreign clinical trials not conducted under an IND were removed not as a direct response to the placebo issue. Instead he said, "What we realized was that a reference in any regulations to a document that we had no control over (such as the Declaration) was not prudent. So we referred instead in the regulations to the ICH good clinical practice guidelines."[41] This realization about the lack of "prudence" in referencing a non-FDA document over which the agency had no control seems to have occurred to the FDA only after the AIDS trials and the 1996 draft of the Declaration, lending some credence to the argument that the Declaration was not a matter of concern to the US agency until after the placebo and related debates erupted.

In the years since the 2008 version of the Declaration, critics of the FDA from the developing world continued to focus on the ties between US policymakers and the pharmaceutical industry. Two

South African medical scientists suggested in 2012 that "the FDA's renouncement of the Declaration may have long-term practical implications and could encourage pharmaceutical companies to take ethical shortcuts in developing and emerging countries."[42] I am not aware of evidence that that has occurred. Others have generalized their critique to an attack on what they see as moral imperialism. This is in some ways the most interesting and yet least resolvable question. Certainly the "optics" of the controversy were not flattering to the FDA. "At worst," wrote an international group of ethicists in the *British Medical Journal*, "[the FDA] is creating the impression that it is more interested in facilitating research than respecting the rights of people who are research participants. This has been variously depicted as entrenching different standards for different parts of the world (ethical pluralism), establishing the U.S.'s right to unique policies (exceptionalism), and one country imposing standards on others (moral imperialism)."[43]

In response to the perception of imperialism in drug regulation, Latin American authors, for example, have introduced the notion of "hard bioethics or intervention bioethics, in support of the interest and the historical rights of the population economically and socially excluded from the international development practice."[44] The background of this hard bioethics movement is an argument about inequalities of international drug trials, including not only the placebo controls issue but also resource allocation well beyond posttrial obligations to local communities. The authors of the "hard bioethics" statement refer to a "10/90" gap in which 90 percent of the economic resources address the health needs of 10 percent of the world population. The revised Declaration of 2008 was followed later that year by a congress of the Latin American and Caribbean Bioethics Network in Cordoba, Argentina. The Declaration of Cordoba excoriated the new Helsinki revisions as unethical and inequitable. In its place, the Cordoba meeting endorsed the UNESCO 2005 Universal Declaration on Bioethics and Human Rights, in particular its article 15, paragraph 1 on benefit sharing: "Benefits resulting from any scientific research and its applications should be shared with society as a whole and within the international community, in particular developing countries."

The controversy about the Declaration seems to be a classic case of protagonists operating on different levels of discourse. To US officials, the matter reflected a disconcerting instability in the successive versions of the Declaration that had not presented itself before, one that centered on a few narrow technical issues that could compromise the most informative possible data from clinical trials. To their critics the regulators' position was a manifestation of the US's proclivity to go it alone, especially if the result happened to benefit a powerful US-based globalizing industry. Though it is hard to see how two such divergent standpoints can be reconciled, it would have been instructive if a forum had been available in which the exponents could have encountered one another.

And in spite of the US position, the WMA continues to reflect on updates to the Declaration and promotes it as a global standard. Whether clinical trials will be strengthened as a consequence of the US position or international ethical standards will be weakened remain open questions. What cannot be doubted is that the pharmaceutical industry's concentration on drug development shaped by market considerations has not ended, nor has the international medical community's resolve to work within the ethical constraints of the realpolitik of clinical trials. In terms of the argument of this book, the article 30 dispute represents conflicts within the rules-based order for bioethics as the discourse extends beyond the consent standard to public health, tensions that were starkly rendered in the COVID-19 pandemic.

CIOMS

The human experiments aspect of bioethics was and remains the focal point of the field as it has crystallized into various institutions within the post–World War II order. Human research ethics was the domain in which the evolution of values in the physician-patient relationship first played out, a conversation that would later extend to the realm of routine medical encounters, public health, and the laboratory. That focus on human experiments would barely shift for decades, and in some ways is still the dominant rationale for bioethics as an institution, but in retrospect one can see that from the very beginning the modern bioethics discourse was on course to expand. An interesting and lesser-known actor in global bioethics is one that signifies what became a field with a more comprehensive mission than human research ethics, one with an initial agenda set in the wake of the doctors' trial but quickly embraced a far more diverse range of topics. The Council of International Organizations of Medical Science (CIOMS) is emblematic of the rules-based order and, within it, the global grounding of bioethics that evolved since the late 1940s.

A UNESCO and WHO partner founded in 1949 to coordinate international medical congresses and to standardize medical nomenclature, CIOMS also relates to the International Federation of Medical Students' Associations, and the International Union of Basic and Clinical Pharmacology. To the layperson these entities may seem obscure, but the web of organizations to which CIOMS relates is critical to the international pharmaceutical industry, signifies the rules-based order within which global medical science operates and, again, exemplifies the ubiquity of bioethics in that system. True to its name, at first CIOMS helped international medical science organizations exchange information and provided financial and material assistance to medical science congresses. From the late 1960s to the late 1990s, CIOMS held round table conferences nearly every year on topics including "the dilemma of human experimentation" (not surprisingly, its first roundtable) in 1967;[45] and "poverty, vulnerability, the value of human life, and the emergence of bioethics" in 1994.[46] In 1990, CIOMS created the ICH—the source of the rules that the FDA substituted for the Declaration of Helsinki—to sponsor dialogue between the pharmaceutical industry and regulatory authorities and to establish guidelines for global drug development. Gradually the organization's focus shifted to bioethics guidance documents and, especially in the twenty-first century, guidance documents on "pharmacovigilance," defined by WHO as "the science and activities relating to the detection, assessment, understanding and prevention of adverse effects or any other medicine/vaccine related problem."[47] A couple of years before the pandemic, CIOMS published guides to vaccine surveillance and safety communication, work intended to guide global cooperation that would have vastly improved the rollout of vaccines and therapies for COVID-19 had the pandemic itself not exposed so many of the cracks in the international system.

Taken together, the Nuremberg Code, the Declaration of Helsinki, and the CIOMS guidelines represent decades of development of certain core bioethics norms. They specify for medical science more general postwar covenants like the Universal Declaration of Human Rights and others on civil, political, economic, social, and cultural rights that also sometimes make reference to human experiments. Also making their appearances were regional declarations that fall somewhere in between treaties and soft law that are thought to bind their signatories, such as the American Declaration of the Rights and Duties of Man (1948), the European Convention on Human Rights (1950), the American Convention on Human Rights (1969), the African Charter on Human and Peoples' Rights and the OAU Resolution of Bioethics (1996), the Council of Europe Convention for the Protection of Human Rights and Dignity of the Human Being with regard to the Application of Biology and Medicine (Oviedo Convention, Spain, 1997), and others. The point of this list is not to drive the reader into an alpha state (though that is a risk), but to observe again how densely populated the universe of the rules-based order has become. Perhaps these regional conventions represent more subtle ideological rebellions against the prominence of American bioethics than the hard bioethics advocates of Latin America, though they seem more like efforts to translate the same broad principles into local terms.

Unlike the Code, which represents a single unique moment in the emergence of an ethics of medicine amid the birthing of a new rules-based order, the other documents were developed over decades through postwar institutions. They were inevitably reflections of the evolving global scene, from the brief postwar respite to the Cold War, to the era of American triumphalism over Soviet Communism, and to the reshuffling of power and alliances in the first quarter of the twenty-first century. Despite doubts about global inequities, they are testimony once again to the power of peer networks in medical science that operate largely in terms of an invisible college along the general lines set out in the postwar human rights and bioethics order, albeit one in which the outsized American role may be celebrated or derided but hardly denied. Indeed, the American way of bioethics, with some significant exceptions, reached well beyond the realm of research with human beings to create a model of medical culture and its attendant values that radiated throughout the rules-based order.

4
The Hegemon at Home

As the rules were being laid down in the new international order, how did the dominant power look to its domestic household, especially in the domain of medical ethics? In August 1946, while Leo Alexander was developing evidence for the trial of the Nazi doctors, a US Public Health Service (PHS) physician named John Cutler arrived in Guatemala City to prepare a treatment program for sexually transmissible diseases (STDs) at the military hospital. STDs had been a major problem for American forces during the war, so improved methods for diagnosing and treating them were a national security imperative. Over the next several years, in cooperation with Guatemalan health authorities, Cutler's group conducted diagnostic studies with children and inoculation experiments with more than a thousand sex workers, soldiers, persons in prison, and patients in psychiatric care. Cutler's PHS supervisor gave him permission to do the studies without consent, developing new tests to detect STDs as well as deliberately giving people syphilis to try out the new wonder drug, penicillin.[1] By the time the intentional exposure experiments ended in 1948, the American judges in Nuremberg had published their ethics code, which would clearly have prohibited these studies.

The Guatemala STD studies were unknown until 2010, when Wellesley College historian Susan Reverby discovered them through John Cutler's papers at the University of Pittsburgh; the subject of her research was Cutler, who also had a leading role in the infamous syphilis study in Tuskegee, Alabama. In 2010, President Obama ordered his bioethics commission to produce an in-depth report, work in which I participated as a senior advisor to the commission, and apologized to the president of Guatemala for this abuse of the country's citizens.[2] Considering that the medical trial and the Guatemala experiments were going on at the same time, the result was, to put it mildly, ironic: The emerging dominant world power was at once eager to shape a new order in accordance with its vision of the rules, while also trying to ensure that it took care to minimize its vulnerabilities in any future conflict, which happened to involve gross violations of the rules being promulgated. Exactly how did this tension arise in the arena of medical science?

National Insecurity

During the years 1946 and 1947, US policymakers were preoccupied with the reconstruction of Europe and Japan, the emerging challenge of Soviet power, and the basic revision of America's position in the world. Hopes were high for a new rules-based order that both avoided the inherent instability of the old imperial system while also establishing a central role for the United States as the global hegemon in a Pax Americana. Operating in a country and a world weary of war and eager for normalcy, military planners nonetheless felt the need to sustain a posture of readiness: a posture that included exploiting whatever advantages science could provide, including medical science. This same national security establishment that wanted to press its new postwar advantages was simultaneously pulled in a somewhat different direction, needing to adhere to the rules it was promulgating in places like Nuremberg. How could it do both?

In a critical setting for the resolution of that question, a few officials and medical researchers within the United States' new civilian atomic energy agency engaged in a tense debate about human research rules that culminated in what they called "informed consent." These developments were precursors of the discussions that began in the late 1960s in the world of civilian medicine, culminating in the work of the National Commission for the Protection of Human Subjects of Biomedical and Behavioral Research (1974–78) and the current federal regulations on human research protections. The National Commission was created by the National Research Act (1974) in the wake of the US Public Health Service syphilis study revelations, resulting finally in the human research rules that have become a cornerstone of US bioethics. In this same period, the US Senate and the Ford administration investigated disclosures of Central Intelligence Agency (CIA) experiments on unsuspecting American citizens with lysergic acid diethylamide (LSD) and other psychoactive drugs.[3] In many ways, the Cold War national security discourse about appropriate human research policies laid the groundwork for the framework of principles classically expressed in the National Commission's *Belmont Report*—respect for persons, beneficence, and justice—which have become a mainstay of US biomedical research policy.

The national security state is a shorthand for the central goal that has driven US policy since the end of World War II: to sustain American military superiority over all actual and potential adversaries. The concept is familiar to Cold War historians and political scientists. The American national security state that was established in the wake of World War II was called into question with the collapse of communism and was fortified by the "global war on terror" (GWOT). This framework has been shaken by the rise of China as a peer competitor intent on creating a new order, in league with a dependent but fossil fuels–rich Russia. In the life sciences, strengthened consent standards for human experiments, though conventionally thought to be antithetical to state interests during an international crisis, can be seen as an attempt by the postwar national security state to protect itself from critics of expanded governmental power and especially to conform to the rules it was promulgating as part of the rules-based order.

The national security state in its historic and current manifestations gains its power precisely from the way it engages the resources of society, such that there is in practice no clear boundary between the civilian and the military worlds. Before World War II, the Yale political scientist Harold Lasswell coined the term "garrison state" to describe a thoroughly militarized nation in a state of continuous war. Conservatives of that era feared that the resulting expansion of state power justified by a chronic wartime–footing would threaten individual liberty and fundamentally alter the underlying values of the American experiment. Some leading liberals like Columbia's John Dewey were also ambivalent about American involvement in World War II due to fears about the long-term consequences of militarization in American life. These fears continued after the war in the form of congressional opposition to President Truman's proposals to reorganize the national security establishment and change budget priories, initiatives that were ultimately successful. In this sense, it is instructive to recall Daniel Yergin's 1977 observations about the concept of national security as it functioned during the Cold War:

> We must remember that "national security" is not a given, not a fact, but a perception, a state of mind. . . . The doctrine is characterized by expansiveness, a tendency to push the subjective boundaries of security outward to more and more areas, to encompass more and more geography and more problems. It demands that the country assume a posture of military preparedness; the whole nation must be on permanent alert. There was new emphasis on technology and armed force.

> Consequent institutional changes occurred. All this leads to a paradox: the growth of American power did not lead to a greater sense of assuredness, but rather to an enlargement of the range of perceived threats that must urgently be confronted.

Science played a critical role in the envisioned national security establishment. World War I had already produced the National Research Council, its remit to bring civilian science into military preparation and the war itself. The importance of the wartime alliance between science and the military for the World War II Allied victory was beyond dispute, not only in the most obvious examples of war-fighting technologies like sonar, improved armaments, and especially the atomic bomb itself, but also in less obvious ways through military medicine. A two-volume report called *Advances in Military Medicine* published after the war attested to the many areas in which medical science enlisted for the war effort had achieved breakthroughs in the basic understanding of medical sciences and in improved therapies, particularly those associated with the infectious diseases and the bacterial disorders that plagued soldiers. Perhaps the most impressive example of the benefits of government support of medical research was the rapid advancement in the development of penicillin, the "wonder drug" whose refinement came too late to be of much use during the war but proved to be a great boon in the postwar era. At first, academic institutions feared the intrusion of the national government in their research priorities. This attitude changed drastically in the 1930s as new academic leaders welcomed government support but tried to minimize interference in scientific work, with mixed success.

No institution exemplified this new cooperative spirit between government and the academy more than MIT, which under its president Karl Compton, devoted itself to the application of physical theory in order to solve a host of engineering challenges. Harvard president James B. Conant, a chemist by training, and MIT vice president Vannevar Bush, an early computer scientist who became president of Washington's Carnegie Institution, led the Roosevelt administration's crash program to realize the benefits of scientific research for the war. Bush led the Office of Scientific Research and Development (OSRD), which infused millions of dollars into scientific research during the war, mainly through the Committee on Medical Research (CMR). The CMR's projects included several that involved human experiments with conscientious objectors, persons in prison, and persons with intellectual or developmental disabilities. Perhaps the most famous of these were the malaria experiments that I described in chapter 2, which the defense lawyers at the Nazi doctors' trial later held it up as an example of Allied hypocrisy, even though it did not take place amid a Holocaust in which death was an acceptable endpoint and genocide the policy goal. (Indeed, the case can be made that there was a degree of voluntariness among the men, despite their incarceration.[4]) The OSRD was also a crucial player in the contentious wartime discussions about whether the nation's atomic energy resources would eventually be under civilian or military control or some combination.

The Plutonium Injections

There were other, still more provocative human experiments during the war. The successor to the "Manhattan Engineer District" (the code name for the atomic bomb development project, later known as the Manhattan Project), was the new civilian Atomic Energy Commission (AEC). The AEC was immediately confronted with an awkward example of the simmering conflict between military and civilian concerns. The Project's deputy medical director proposed that a report about secret plutonium injections be declassified. These injections had taken place at several major research hospitals from 1945 to 1947. The scientific considerations that favored sharing data were received skeptically by an AEC declassification officer, who wrote on February 28, 1947, that this particular report,

> appears to be the most dangerous [among a number being reviewed] since it describes experiments performed on human subjects, including the actual injection of the metal plutonium into the human body. . . . Unless, of course, the legal aspects were covered by the necessary documents [perhaps a reference to patient consent forms], the experimenters and the employing agencies, including the U.S., have been laid open to a devastating lawsuit which would, through its attendant publicity, have far-reaching results.

In addition, "the coldly scientific manner in which the results are tabulated and discussed would have a very poor effect upon the public."[5] A few weeks later the AEC's medical director stated his agreement with this recommendation.

Thus, the outcome of this early contest between scientific openness and the secrecy needs of national security was determined by the requirement for the emerging national security state to be spared the public embarrassment that would come with lawsuits and scandal. AEC officials were aware of the trial of Nazi doctors taking place at Nuremberg at exactly the same time they were dealing with the plutonium experiments declassification issue. As the implications of this incident became apparent, they sought to protect the government from the risk of subsequent embarrassment about human experiments by requiring what they called "informed consent" of any participants in future agency-sponsored radiation experiments, the first known use of this term in writing. Various interests competed for attention in this episode: the state's interest in avoiding liability and damaging publicity and medical scientists' interest in protecting their professional autonomy and authority. Later in the 1940s, when the AEC and the Pentagon were jointly considering the development of nuclear-powered aircraft, questions were raised about how to determine that the power source would not impair the crew; the risk that one of the planes might fall from the sky and create a radiation hazard was enough to end the project. Human experiments were desirable, but the most likely population, long-term prisoners, invited disconcerting comparisons to the Nazi crimes.[6]

Despite scattered contemporary references to the trial, the immediate influence on American medicine of the Nuremberg Code developed by the Nazi doctors' judges was minimal, but government officials were sensitive to both its legal and political implications. The Code surfaced in military medical planning as a direct result of President Truman's creation of a single Department of Defense (DoD) from the departments of the Navy and War. Undertaken in 1949, this massive bureaucratic overhaul predictably left numerous gaps in the new department's policy framework, including the field of human experimentation. There were fears that the Soviets were outstripping US military scientists in the development of unconventional weapons and in the defense against them, including the conduct of human experiments. From 1950 to 1953 several internal Pentagon advisory committees deliberated on the matter of suitable guidance. In the end, both the military and medical members of all these panels largely agreed that a written policy was a bad idea, preferring instead to rely on an unwritten code of ethics and the virtues of those in charge rather than opening the matter up to legal scrutiny.[7]

At first, like the AEC leadership, the Pentagon's civilian leadership was reluctant to assert itself in what was turning out to be a controversial area. Truman's secretary of defense, Robert Lovett, had his people working the problem on a track parallel to the military medical advisory process. The job of finding an acceptable policy was handed over to Defense Department legal counsel, which identified the Nuremberg Code as prevailing international law. A key player in the decision to adopt the code was Lovett's deputy, Anna Rosenberg, assistant defense secretary in charge of manpower. She took the proposed use of the Nuremberg principles a step further by adding a written consent requirement, a consequence perhaps of her experience as a leading labor-management specialist in New York: Contracts are only concluded when they are reduced to writing.

In spite of Rosenberg's advocacy and Lovett's sympathy for the proposal, internal opposition to any written policy at all was so strong that the civilians in Truman's Defense Department decided it would be best to refer the matter to the incoming administration of Dwight D. Eisenhower, as the new defense officials would have to live with the consequences of imposing an unpopular policy on the military medical bureaucracy. The matter was still urgent, the outgoing officials explained to their successors, because there was no departmental rule that would allow important experiments on the human effects of atomic, biological, and chemical weapons. There is reason to believe that Lovett himself briefed Charles E. Wilson on the issue as one among many then facing the department. Lovett believed that the 1949 reorganization of the defense establishment needed more work, so he was involved in the transition to an exceptional degree.

Finally, the new defense secretary, Wilson signed off on a top-secret memorandum to Army, Navy, and Air Force secretaries on February 26, 1953, making the Nuremberg Code the Pentagon's policy. Secretary Wilson had been chief executive officer of General Electric and, at the time he took office, General Motors. Like Anna Rosenberg, he was experienced in labor negotiations and might well have approached the human experiments issue in a matter-of-fact manner as a contract problem, a deal between the department and the experimenter on the one hand and the human volunteer on the other. This would have been consistent with the management style Eisenhower wanted him to bring to the job, running the DoD like an industrial corporation. Wilson was also charged with implementing Eisenhower's "New Look" defense strategy, which included greater reliance on nuclear weapons. Any help the medical experts could give on human factors in atomic warfare would clearly be welcome. In 1953, amid simmering conservative skepticism about the growing empowerment of a security state, medical and military reservations about participant consent were of less concern to policy makers than the need for legitimizing the new order. What the AEC's leadership could not accomplish in 1947 the new Eisenhower Defense Department, with a formidable ex-general in the White House, could.

In January 1953, at the same time Lovett's staff was communicating with the new defense officials about the need for a written participants consent policy, a depressed psychiatric patient died of an overdose of a mescaline derivative in a secret experiment at New York's Psychiatric Institute. The sponsor was the Army Chemical Corps, which engineered a cover-up in cooperation with the New York State government. It is not known whether this incident further impelled Secretary Wilson to sign off on the policy without delay, but he would likely have been informed that the Army had sponsored research in which a civilian death had occurred. Only three months before, in October 1952, the Chemical Corps raised questions with DoD legal counsel about release forms for "human guinea pigs" at the same time another prisoner died in an Armed Forces Epidemiological Board–sponsored hepatitis study.[8]

That defense officials would want to introduce a clear policy to protect the government from human experiment scandals may surprise those who suppose that expanded state power led to Cold War practices that were antithetical to the high ethical principles of the Nuremberg Code, such as the CIA and Army's experiments with hallucinogenic substances. It might seem odd that the state sought to protect itself from scandal by advancing participants consent, rather than imposing its will on innocent human beings. But seen through the lens of the garrison state, a more complex analysis is required. In fact, both scenarios played out at different times and in different circumstances during the Cold War. Each was justifiable according to the needs of the national security state, which sometimes was concerned about its perceived legitimacy at home, at other times concerned more with containing aggressive Soviet intentions. Thus, it is noteworthy that during the middle and later 1950s some of the most egregious abuses of human beings in psychoactive drug studies occurred not in the DoD but in the CIA, an entity that was a direct result of the national security reorganization and less subject to layers of scrutiny than the military.

Beecher

In the later 1950s, the Nuremberg Code–based policy withered as it was variously interpreted and sporadically applied. Functionally, the national security state had less need to protect itself from the criticism that put traditional American individualism and anti-state sentiment at risk. The pre–World War II conservatives and isolationists lost their clout as that generation faded and as the Korean War abroad and Communist scares at home rattled Americans about their safety from alien threats. As embodied in the 1960 presidential candidacy of Richard Nixon (his rival, Democrat John F. Kennedy ran hard to Nixon's right on this point), enhanced state power was no longer seen as antithetical to American values but was considered central to their survival in the competition with the Soviet Union. In that respect, neither national party was a devotee of small government. Although the ethics of human experiments was not a top-of-mind issue for most policymakers, the question of the limits of state powers was quite pertinent in federal agencies that, in pursuing their missions, wanted answers to questions about medical science.

In the early 1960s the uniformed services created written policies on human experiments, beginning with the Army Regulation 70–25, which contained vestiges of the Nuremberg Code's language. At this time, the services were engaged in more regular publication of all policies. By then the vigorous internal Pentagon debates about standards for human experiments of the late 1940s and early 1950s were all but forgotten. The only significant opposition seemed to come from academic scientists whose institutions contracted with the Pentagon and who objected to being brought under an external set of rules, especially one that had been devised in light of the Nazi atrocities.

One institution that protested loudly was Harvard, which extracted a concession from the Army surgeon general that the "principles" in the Army contract were "guidelines" rather than "rigid rules."[9] Among the Harvard scientists who joined the successful opposition to the Army regulations was a future icon of bioethics, Henry Beecher. His landmark 1966 paper in the *New England Journal of Medicine* on human research ethics did more to advance the issue in the United States than the trial of the Nazi doctors. Scott Podolsky found that a year before the Harvard protest, Beecher was among those who wrote to Senator Estes Kefauver about certain additions to the Federal Food and Drug Act then being considered, objecting to the adoption of an informed consent requirement for clinical research.[10] Clearly Beecher was no admirer of medical ethics rules laid down outside the medical profession. He preferred to rely on the virtuous medical investigator, who is in any case the ultimate guarantor of ethical science, hence his opposition to the Army's Nuremberg Code–based policy being imposed on academics.

What was the source of Beecher's strong commitment to the ethics of human research? One element was likely his own Army-sponsored research in the early 1950s, in which he directed projects that exposed healthy participants to legal hallucinogens without their consent, in particular LSD. As Beecher himself never discussed this experience, it was left to his former research fellow, Louis Lasagna, to describe the work in an interview with the Advisory Committee on Human Radiation Experiments in 1994. So it was that research sponsored by the national security state inspired one of its own investigators to take on the task of reforming the system of human experimentation.

Apart from the Harvard protest, how can we account for the apparent lack of opposition within the defense establishment when the research rules were published, as compared to a decade before? A set of exemptions in the 1962 rules may be the explanation. These exemptions stated that the written consent requirement did not apply if the intent was to benefit the subject, when the studies concerned hazards of training exercises or other normal duties, or when disclosure would compromise the value of the data to "human factors," presumably a reference to unblinding the study's goals or subject randomization. The second exemption, on training exercises, institutionalized what had been a problem with the Wilson policy throughout the 1950s. As I have elaborated in my book *Undue Risk*, and as the Army inspector general concluded in 1975, the rules were applied quite haphazardly. In particular, activities that involved medical monitoring, such as the deployment of over two hundred thousand men to atomic test shots from 1953 to 1962, were often not called human experiments but training exercises. Virtually any activity related to readiness would qualify under that rubric, giving the military wide discretion over potential harms to personnel. When the case of a soldier who had been in LSD experiments reached the US Supreme Court in 1987, a 5–4 majority in *United States v. Stanley* ruled that the veteran could not sue for injuries alleged to have been incurred in the service because to do so would bring the courts into military matters and compromise discipline, overcoming a minority that cited the Nuremberg Code. As a result, the Wilson policy only applied, in effect, on those occasions in which the medical research aim was determined by relevant decision makers to be the central rationale for the activity. As neither the military nor medical establishments of the 1950s were familiar with the concept of written participant consent, there was much inertia in this regard, and little incentive for those in the command structure to impose consent requirements in specific cases.

Although written consent was the primary focus of controversy in the 1950s, the US government medical establishment introduced formidable innovation that transformed the human research system, a bureaucratic entity called an Institutional Review Board (IRB), a label that was just obscure enough to deflect some of the ire of investigators concerned about their professional discretion. Called research ethics committees (RECs) elsewhere, IRBs are charged with the prior review of research protocols, including consent arrangements. At the federal level they were the brainchild of a politically astute NIH director named James Shannon. Worried about blowback to the NIH of early 1960s research scandals, Shannon responded by insisting that the expert review process the NIH research hospital had created before its 1953 opening should be required for all research institutions receiving federal funds. In 1966, the US Surgeon General ensured that this rule would be adopted, a few months before Henry Beecher's influential paper was published. As Laura Stark writes, "The invention of bioethics as a profession and the invention of expert review are two parallel stories with one common cause: medical researchers' concerns over their legal liability in clinical studies and clinical care."[11] The IRB model quickly became part of the global clinical trials system, another decisive moment in the dominant role of US bioethics.

SHAD

For roughly twenty years after the Wilson memorandum, the more the activities in question were considered training exercises, and the less they resembled a clinical intervention, the more likely it was that consent would not be sought. There were some significant exceptions, such as the "panic" studies of a few participants in the atomic test shots (mainly officers who did give written consent), and the Defense Department–sponsored total body irradiation studies at the University of Cincinnati (where researchers seem not to have been told about the written consent requirements), but the generalization holds. An interesting example that was not widely known, even following the massive declassification associated with the work of the Advisory Committee on Human Radiation Experiments in 1994–95, is Project SHAD, for "Shipboard Hazard and Defense."

Following persistent complaints by 1960s veterans, a preliminary Pentagon investigation in 2002 confirmed that dozens of these tests took place. Although it had long been known that environmental tests of biological and chemical simulants took place during the Cold War in the airspace above major cities and in the New York City subway system, the Project SHAD revelations pointed to exercises on the oceans. Sailors were exposed to aerosols from tugboats mounted with biological disseminators, from spray tanks mounted on attack aircraft, towers, and other means. The purpose was to assess the vulnerability of naval ships to chemical and biological agents. This case illustrates how a single research program can succeed in meeting ethics standards in some respects and fail in others. For example, it was at first thought that the substances to which sailors were exposed were inert simulates, but the DoD found that sometimes actual agents were used. On these occasions, such as a 1964 Pacific Ocean test called Flower Drum, the DoD found that there were extensive protective measures, including masks. In the SHAD tests, there was some level of knowledge among the men that they were part of an experiment, but there was no consent process of any kind, written or otherwise. In June 2003, the Pentagon released what it considered to be a final set of reports on exercises that might have exposed 5,842 people to chemical or biological weapons.

As we have seen, military personnel have long presented an ethics puzzle about risk exposure: Should they be considered healthy, normal research participants or as participants in a line-of-duty exercise? One of the SHAD tests, called Autumn Gold, took place in the Pacific Ocean in May 1963. A fact sheet released as part of the Pentagon investigation reported that, "According to the Autumn Gold test plan and final report, the crews who participated in the tests were not test subjects, but test conductors."[12] This is a nifty legalism if ever there was one. Again, rules concerning consent to medical research participation would not apply under that classification. Nonetheless, in its review the Pentagon stated that "participants should have been fully informed of the details of each test," and that the training the men received in preparation for the test should have reinforced their understanding. So, on one hand the participants are described as conductors of the tests rather than participants or subjects; on the other, the DoD expresses regret at the lack of informed consent.[13]

Media reports of the syphilis study compelled both congressional and intra-agency investigations. In its wake, along with the Watergate scandal and the national reaction to the chaotic conclusion to the Vietnam War, public mistrust of the federal government reached historic proportions, perhaps only exceeded in the 1850s and the 2020s. Beginning in the late 1960s, and coming into full force in the 1990s, international conventions against chemical, biological, and toxin weapons also militated against offensive research. Human research protections, both in the civilian and military sectors, became far more restrictive and accountable. As the pressures of confrontation with the Soviet Union lapsed into memory, there was also rising interest in learning about Cold War secrets, including the human experiments disclosed as part of President Clinton's Advisory Committee on Human Radiation Experiments. Before the turn of the century, what appeared to be diminishing threats from abroad opened a window to reconsider past acts that might have violated ethics conventions. Without the threat of global Communism was the garrison state and all of its apparatus relevant? Couldn't the country now come to terms more fully with excesses across the Cold War national security establishment and not only those of the CIA that had already been disclosed in the 1970s? Government openness, including the declassification of thousands of documents, gained a modest foothold as the rationale for the national security state appeared implicitly to be facing its greatest challenge in fifty years.

That turned out to be a brief respite.

A New Threat

In the months following the terrorist attacks of September 11, 2001, and the anthrax letter attacks a few weeks later, the George W. Bush administration assessed the new circumstances of a war on terror that would rival the Cold War in the national will required to be victorious. Instead of dealing with nation states, the United States would be at war with shadowy groups, often fluid in nature, motivated by a distorted Islamist ideology and only sometimes in association with established governments. Their methods would be unconventional terror devices intended to disrupt civil society rather than to conquer it with large-scale military means. The new national security policy, announced in September 2002, put unprecedented emphasis on preemptive military engagement as a necessary tactic to root out terrorism before it could strike.

This posture, which seemed to be required for a GWOT, had implications for experiments related to national security. Government critics reacted with concern when, on September 4, 2001, the Defense Intelligence Agency (DIA) acknowledged that it planned to reproduce a strain of anthrax bacterium suspected of still being held by Russia, one that may have been developed during the waning days of the Cold War. The same report indicated that the CIA, as part of a project called "Clear Vision," was building replicas of bomblets also believed to be developed by the former Soviet Union. The bomblets, about the size of a grapefruit, were designed to disperse biological agents, including anthrax, without destroying the microbes in the process. Observers wondered whether these activities could be considered violations of the Biological and Toxin Weapons Convention of 1975. But government attorneys were confident that, because these projects were designed to develop defensive measures, they were not in violation of any treaties. Had 9/11 not occurred a week after the DIA's anthrax plan was announced that might have been a big story; as it was, the country responded instead to a profound sense of vulnerability.

Smallpox is one of the most lethal diseases in human history. On November 13, 2001, Saint Louis University published a news release concerning its NIH-sponsored trial of dilute smallpox vaccine, one that had begun before 9/11. Saint Louis University was only one of several academic centers that participated in the study. The trial was said to be proceeding as planned but that recruitment in the desired age range, 18 to 25, was the major delay. But public anxiety about the anthrax attacks was such that there was no longer any difficulty recruiting volunteers. Ultimately the trial was successfully completed, indicating that immunity was achieved in dilutions of the original vaccine up to ten times. Less than a year later, federal officials confirmed plans to test the smallpox vaccine with children at Cincinnati's Children's Hospital Medical Center and at Harbor-UCLA Medical Center. The stated goal was to ascertain the optimal dosage in children in the event a massive inoculation is needed following a bioterrorist attack. The trial was never carried out, but that such experiments with children could be seriously contemplated was a measure of the times.

"Batshit Crazy"

No account of bioethics in the post-9/11 era can be complete without a discussion of the treatment of "detainees," particularly in Guantanamo. As if the images of humiliated prisoners after the invasions of Iraq and Afghanistan at US black sites weren't enough, the Bush administration endorsed "enhanced interrogation" techniques like waterboarding that had long been considered illegal under international law. The Tokyo war crimes trials, technically the International Medical Tribunal for the Far East, included the "water treatment" among the crimes committed by Imperial Japanese leaders. Unlike the Japanese prisoners, no formal charges against the detainees in Guantanamo were ever drawn up; within a few years, former Bush administration officials admitted that innocent people were among those detained.[14] But even if all turned out to be guilty of participating in attacks on Americans that would not justify either the acts of torture or the failure by American authorities to respect due process of law. In the words of the introduction to the popular TV show Homeland, the country went "batshit crazy" after September 11.

That observation understates the case. More than twenty years after the opening of the camp in 2002, thirty prisoners remained. By then nine had died in the camp. In 2023, the first UN inspector allowed in the camp denounced the "cruel, inhuman, and degrading treatment" at the prison, "the cumulative and intersectional harms arising from continued detention, deep psychological distress, deprivation of physical, social, and emotional support from family and community while living in a detention environment without trial for some and without charge for others for 21years, hunger striking and force-feeding, self-harm and suicidal ideation, and accelerated aging."[15]

I had been following conditions at Guantanamo since 2007, when the Defense Department's assistant secretary for health, S. Ward Casscells, asked a few bioethicists to be part of an informal discussion group about the issues at the prison. This was a good faith effort by Casscells, a career Army officer who was trained as a cardiologist. Newly appointed, he brought a new attitude to the issues, reaching out to critics in light of the outrage over the treatment of the detainees and the Department of Justice's notorious 2003 memo that authorized the CIA's "enhanced interrogation" program.[16] The techniques that amounted to torture were quite different from Army manual FM 34–52 (1992) that specifies psychological techniques oriented toward developing rapport and prohibits threats of physical violence, let alone physical abuse. The role of psychologists and the American Psychological Association in designing this torture program has been well-documented.[17]

I participated in several of these "back channel" sessions in Washington, where we had the opportunity to talk with Casscells and other civilian department officials. In one session we spoke for several hours with the Army general in command of the base. Besides the ethical issues concerning interrogation, which had by then become matters of great public concern, especially with the highly critical public statements of former interrogators themselves, we also had a lengthy discussion about force feeding. Some detainees had gone on hunger strikes as one of the few means of protest available to them. Casscells, who died in 2012, seemingly got no traction from others in the Pentagon who did not welcome his efforts.

This episode and others should have taught me not to be surprised when bioethics issues crop up in the national security setting. I was surprised, however, when a little-noticed CIA draft memo, apparently written near the end of the George W. Bush administration in 2008, expressed annoyance about bioethicists' criticism of the interrogation program. Released in 2018 following a Freedom of Information Act request by the American Civil Liberties Union, the agency's Office of Medical Services (OMS) eighty-eight-page draft was entitled "Summary and Reflections of Chief of Medical Services on OMS Participation in the RDI Program," where RDI stands for rendition, detention, and interrogation. The heavily redacted, formerly top secret, draft document is a vigorous defense of the agency's actions in response to the threat posed by Al Qaeda in the aftermath of 9/11 and the uncertainty associated with the anthrax attacks. A climactic eight-page section on "ethics" makes it clear that the medical office leadership was deeply offended by published criticism by bioethicists about its moral compass, especially the several that appeared in high-profile publications like the *New England Journal of Medicine*: "The thrust of these articles . . . was that there was little or no place for medical personnel or psychologists in interrogations, and especially those involving coercive techniques or designed with medical input on detainee vulnerabilities." The draft then recites a litany of defenses of the RDI program: that "OMS personnel long since had resolved personal ethical concerns by the time such commentaries appeared," that there were no penalties for those who refused to participate, that "actual application of enhanced techniques was much more modest than the press image," and that "these techniques had been dramatically successful in producing indispensable intelligence not otherwise obtainable." None of these assertions have worn well with time.

More interesting in the domain of bioethics is the draft's systematic defensive review of elements of the rules-based order. The memo argues that the ethicists' appeals to Common Article 3 of the Geneva Conventions did not apply because the administration had determined that the detainees were not legal combatants. Umbrage is then thoroughly unleashed: "Further, along with railing at the Administration's permissive interpretations and asserting a humanitarian obligation to follow the Geneva Accords even if they were not legally binding, ethicists turned to another potentially valuable ally to carry their case—the professional associations of organized medicine." The first ethicists' ally cited is the WMA, which in its 1975 Declaration of Tokyo opposed physician participation in torture. The next ally cited is the American Psychological Association's "more nuanced" 2005 statement about psychologists' ethical obligations in "national security-related work" as well as the American Psychiatric Association's "more restrictive" statement. As to the ICRC's "formulaic view" about patients' best interests and dignity, the OMS draft expressed confidence that in response to the hunger strikes such principles were honored because "excepting only a handful of involuntary feedings, consent was obtained before all medical procedures or they were not undertaken." Through all this, what seemed especially worrisome to the agency was the possibility that state licensing boards would revoke the licenses of professionals involved in the RDI program.[18]

In 2010, a couple of years after the OMS memo was drafted, Physicians for Human Rights (PHR) published a report alleging that after 9/11 the CIA had "conducted human research and experimentation on prisoners in US custody as part of this monitoring role," a reference to the medical monitoring of every instance of "enhanced" interrogation. The PHR white paper drew on publicly available documents to draw its inferences about the systematic nature of the information collection, noting that such activities would be in violation of the federal regulations governing research involving human participants. These novel measures were said to include new methods for waterboarding, pain susceptibility, and varied application of sleep deprivation: "The outcomes were used to monitor stress levels in individual detainees ongoing 'enhanced interrogation' but also, apparently to perform research with the goal of calibrating interrogation technique in the interest of achieving maximum effect with detainees in the future. In short, OMS had conducted human subject research," according to the report. But the reception to the allegations in the PHR report got mired in the question whether any such information collected truly meets the definition of research as "generalizable knowledge" or was more similar to routine program quality improvement, even if the program it sought to improve was execrable.[19]

As Jonathan Marks has noted, the US approach to the legal issues concerning the treatment of the detainees in Guantanamo was essentially to define them out of existence, such that Common Article 3 of the Geneva Conventions did not apply to them. Following 9/11 another leading protagonist in the rules-based order, the United Kingdom, took a different approach. Addressing the indefinite detention of suspected international terrorists, the Blair government found an exception to prevailing human rights conventions due to a "public emergency threatening the life of the nation." As Marks observes, "there is an exceptionalism in both the U.S. and the U.K. governments' approaches. But the U.S. approach seems to me more contemptuous of the legal order that it co-created."[20]

Classified Research

In these incidents we can see how, specifically in the bioethics context, the US was caught between the rules-based order it helped to promulgate, and on which it depended for its hegemonic power, and challenges to its maintenance of that power. The PHR's allegations about the RDI program point out additional issues related to research involving human participants that is classified. Currently the federal government reports that there is no classified human research taking place (as distinct from potential *applications* of the research that might be used against an adversary), but it is not difficult to imagine circumstances in which such activities would be thought important so that terrorist groups would be denied information about American capabilities. Yet, there are no rules in place in the federal regulations governing human research that create a setting for classified studies. That itself is an invitation to game the system by calling the work "program evaluation," or "field testing," or some other such terminology, even though it could be seen as meeting the criteria for generalizable knowledge.

Reform is needed, perhaps in the form of an addition to the federal rules, a subpart F that would specify the special requirements for informed consent and prior review by an IRB. To satisfy the standard criteria of valid informed consent, a new subpart would require security clearances for all parties, including the IRB members and the human volunteers themselves. Because there would be no publicity with regard to IRB decisions, the standard check of public scrutiny would not be available. To compensate for the lack of public scrutiny, at least one member of the classified IRB should not be a federal employee and there should be a method of appeal of an IRB decision by any member to a superior authority.

These essential elements were included in a presidential letter signed by President Clinton in 1996, which still technically applies to all seventeen federal agencies that conduct or sponsor human experiments, including those with classification authority. The presidential letter was stimulated by recommendations of the Advisory Committee on Human Radiation Experiments in its 1995 final report, and it was to have been transformed into a common rule amendment. This complicated amendment process, which would involve all seventeen common rule agencies, has still not been completed. But even these IRB-oriented measures may not be enough. Deviations from protocol are a familiar problem in the unclassified research world and have been implicated in the deaths of research participants. Some equivalent to the function of data safety monitoring boards or committees, normally appointed to oversee double-blinded efficacy trials, should be part of any comprehensive plan to protect individuals who agree to participate in classified research studies. Members of these boards should also be given suitable security clearances and include nongovernment employees. On pain of inconsistency, the rules-based order begins at home.

5
Bioethics at Large

The origin story of the modern field known as bioethics, and its relationship to traditional medical ethics, is more complicated than the American story alone. Despite the faded memories of the European medical ethics legacies, the typical American narrative of the origins of bioethics in suburban New York and the Georgetown neighborhood of Washington, DC must obviously be qualified. Nonetheless, the American influence over the system of international drug and device trials that was formalized in the 1950s and 1960s can hardly be denied, as we have seen in the Helsinki story. This globalized system includes the institutionalization of what came to be the requirement of informed consent by the human volunteers themselves or their legally authorized surrogates, a requirement founded on the sentiment behind the emerging principle of respect for autonomy. This philosophical shift away from medical discretion and toward the patient decision maker is now embedded in the money and power of the international pharmaceutical industry. In this respect, what began as a philosophical trend is now institutionalized as a regulatory requirement. Even a critic of an excessively individualistic and "Western" application of the principle of respect for autonomy, while arguing that "the informed-consent process cannot simply be transferred from developed to developing countries," does not deny that investigators are nonetheless bound by an "informed consent transaction."[1]

Bio-Ethics to Bioethics

Rather than having been mapped originally onto a more patient-centered view of medical ethics, the neologism seems to have been coined by the German Protestant pastor Fritz Jahr in his 1927 article, "Bio-Ethics: A Review of the Ethical Relationships of Humans to Animals and Plants."[2] In the nineteenth century, German life scientists sought to overcome Cartesian dualism by investigating environmental sensing mechanisms in biological organisms. A resident of Halle an der Salle, Jahr argued in his *Kosmos* paper that versions of these mechanisms characterize humans, animals, and plants. Echoing Kantian moral philosophy, Jahr proposed the rule: "Respect every living being on principle as a goal in itself and treat it, if possible, as such!" Jahr insists that this rule does not imply a Buddhist injunction against the killing of all animals, but it does prohibit gratuitous suffering. Even the careless picking of flowers was prohibited when, shortly thereafter, they'd be thoughtlessly thrown away.[3]

Writing in the late 1960s against a background of growing environmental activism, especially Rachel Carson's best-selling broadside against the wide use of pesticides in *The Silent Spring* (1962), the University of Wisconsin cancer biologist Van Rensselaer Potter signaled a similar but more urgent call for a deeper respect for, and engagement with, nature. Potter's refrain also reflected Malthusian anxieties about the global food supply that has since been tempered by "the green revolution," at least temporarily. Nonetheless, Potter's call for a "bioethics" reflected the growing sense that a prerequisite for human survival must be actions and policies founded on an understanding of biology. Potter's more academic reflections ran parallel to the ecological activism that came to be embodied in the first Earth Day in

1970, the same year as Potter's paper describing bioethics as a "science of survival."[4]

American-style bioethics became a dominant force in the early 1970s, as the term itself came to be identified more with medicine and the medical sciences than with the broader sense of the human relationship with the natural world. As well, academic centers with a more classic deontological orientation to medical ethics, like those in Eastern Europe, became more isolated during the Cold War; by the late twentieth century, the efforts of the East European deontologists were all but forgotten. In turn, the ways that bioethics has become manifest in the international rules-based order are strongly influenced by the concepts and conversations that developed largely in the US in the 1960s and 1970s. So far, we have seen how this played out in research with human participants. Another question is how reforms in medical ethics standards evolved from the research setting to clinical practice. Research involving human beings became more formalized after World War II, culminating in the emergence of the randomized controlled clinical trial as the "gold standard" of validity, marking it off from the domain of regular clinical practice. Where once there were scattered requirements or guidelines in some countries for consent to being part of a human experiment, consent to research participation was a core feature of the newly conceptualized systematic research process in the expanding international medical science community. Consent to therapeutic interventions outside the research process has a far less definitive postwar history. Nonetheless, modern bioethics has come to incorporate a consent standard in regular medical practice as well as research.

Trickle Down

The essential rationale for a consent standard in the research setting is that, so long as the risks to the participant are acceptable, interventions aimed at the growth of medical knowledge are justifiable and even praiseworthy. But as research consent became more routine in the 1960s and 1970s, it became awkward to maintain that a human being in one context is not entitled to a reasonably similar form of respect in another, regardless of the primary goal. One could say that a consent standard trickled down from research to therapy. Granted that the discourse about clinical consent did not have the same intensity as in the research setting in the decades following World War II, especially when it involved "normal, healthy volunteers," the consent requirement for a proposed therapeutic intervention does have a formal standing in the rules-based order. Article Six of the Universal Declaration of Bioethics and Human Rights (2005) stated that "any preventive, diagnostic and therapeutic medical intervention is only to be carried out with the prior, free and informed consent of the person concerned, based on adequate information. The consent should, where appropriate, be express and may be withdrawn by the person concerned at any time and for any reason without disadvantage or prejudice."[5]

It is not easy to know to what degree the sentiment expressed in Article Six has been manifest in the real world. With few exceptions, such as those involving malpractice claims or fraudulent billing, the clinical encounter is still far less scrutinized than is the research setting. One of the few attempts to compare the prevalence of clinical consent in a number of countries was a 2018 paper by several British orthopedic surgeons; their conclusions were not limited to surgery. In their review the authors asked "how risks and benefits are discussed with patients, and how we can use them to inform our own patients better." Confining themselves to English-speaking countries for ease of comparison, they reviewed case law and clinical guidelines in the United Kingdom, Australia, New Zealand, Canada, and the United States, looking back over sixty years of legal doctrine and precedent. They observed that, where once the test of negligent consent was whether "medics" would have disclosed the same information about risks to the patient, there has been a "shift of focus in consent from the physician to the patient."[6] The "professional practice" standard was articulated in the British courts in 1957. Called the Bolam Test after the patient plaintiff, this case set the standard as conduct that would be supported by a reasonable body of medical opinion. Bringing their own perspective, these British surgeons confirmed the shift of focus that has been observed in the bioethics literature for many years. The professional practice standard of disclosure gradually came under more pressure earlier in the US than in the UK, as in case of *Canterbury v. Spence* (1972), when the "patient standard" was articulated, even though the outcome was not in the plaintiff's favor. Under the patient-centered approach the issue is what that individual patient would have wanted to know about the risks of surgery, not simply what most professionals would have disclosed. A 2015 UK case, *Montgomery v. Lanarkshire Health Board,* reinforced this trend. The court in *Montgomery* found that the standard for the disclosure of an intervention's risk depended finally on whether the risk would be material to a reasonable person in the patient's position.

Consent in the Clinic

Attitudes toward consent to therapy lagged behind the research setting even though the consent standard is implicit in the human rights dimension of the rules-based order. One reason for the lagging response may be that the moral grounds for a consent standard in research with human participants, as contrasted with consent in routine clinical practice, seems like a more obvious ethical requirement than consent in the clinical setting. In the latter, the patient's well-being is supposed to be paramount, not the expansion of scientific knowledge. (One can't dismiss a number of other factors that have come to distinguish the two settings, like the greater legal scrutiny given research than routine practice, the public scandals about human experiments during and since the 1960s, and fears of malpractice litigation.) In its famous *Belmont Report* (1979), the National Commission for the Protection of Human Subjects of Biomedical and Behavioral Research was careful to specify that the reason that the Commission placed its emphasis on consent to research was their limited charge from Congress "to know what activities ought to undergo review for the protection of human subjects of research," without depreciating the importance of consent more generally, especially in cases where an experiment is not technically "research" but still deviates from a recognized medical treatment.[7]

Despite the fact that the main driver of postwar medical ethics since the medical trial has been consent in research, the importance of consent in regular clinical practice has come to be widely accepted. By the time the US National Commission was doing its work in the mid-1970s, in Western countries in particular, consent was seen as ethically important in regular physician-patient relations. An immense change in professional attitudes had taken place in the previous two decades. The evidence for this is assertion can be established by linking a wide a wide range of sources. A 1956 textbook titled *Medical Ethics*, written by Edwin Healy, a Jesuit and professor at Loyola University in Chicago, included a case of a woman with a chondrosarcoma whose physician decided not to disclose her terminal cancer. In an implicit appeal to the therapeutic exception, Healy's analysis and conclusion found the physician's withholding of the truth to be justified so as not to cause her needless emotional distress. This was by no means a unique view. A survey of physicians published in JAMA in 1961 asked how commonly they disclosed a cancer diagnosis. It found that "whether or not physicians tell their patients they have cancer. . . . [T]here is a strong and general tendency to withhold this information. Almost 90 percent of the group is within this half of the scale. Indeed, a majority tell only very rarely, if ever," although most of those surveyed said that if they were in that position they would want to be told.[8] A follow up survey in 1979, also published in *JAMA*, found that "98% of those surveyed reported a policy of disclosure to cancer patients."[9] The sea change in attitudes needs to be taken with a grain of salt: Surveys don't necessarily reflect actual practice, but at least these respondents knew what they were supposed to say, and that change is remarkable.

Thus, what began as a discourse about consent in the research domain, one that I have argued is firmly implanted in the rules-based order, migrated to the clinical domain. (At least in the developed world; there is far less information about what was going on elsewhere, though it seems that as medical cultures modernize, the emphasis on consent tends to increase.) As I have noted, many reasons may be offered for this development, from the growth of civil rights movements to improved clinical medical science with more predictable results. Legal liability and an increase in malpractice litigation might also be cited, but this confuses the cause with the effect: The failure to obtain informed consent became societally objectionable, setting the stage for an appealing legal argument. But the legal and philosophical case for consent to therapy only goes so far. Negligent consent practices in the regular clinical setting rarely result in a financial award to the plaintiff in the absence of an injury.[10]

Whatever the reasons for the shift in attitudes in the latter third of the twentieth century, the traditional Hippocratic role in which the physician would mainly diagnose, prognosticate, and comfort, gave way to more evidence-driven practice with some interventions that could actually make a difference. Paradoxically, as the power to affect an improvement in the patient's condition increased, the physician's historic priestly Aesculapian authority decreased, giving way to worries that doctors were becoming mere "body mechanics," losing their interpersonal skills and traditional roles as confidant and counselor. Gradually, the balance of power between physicians and patients began to shift. Paternalism became a dirty word, autonomy nearly always laudable. By the turn of the century access to medical information online via internet search engines, albeit often inappropriate, misleading, or simply incorrect, was an element of this shifting dynamic. By the late 1990s, some patients began presenting their doctors with printouts from the web, asking for this or that treatment for a condition they believed they had self-diagnosed.

First Among Equals

If American bioethics has been an exceptionally influential force in the global movement, the principle of respect for autonomy has been the tip of the spear. Like the Nuremberg Code and the Helsinki Declaration, the story I have told about the migration of proconsent attitudes from systematic human research to the consulting room bears distinctly Western, and especially American, fingerprints. From the Nuremberg Code on, the telltale sign is an implicit primary appeal to a principle of respect for autonomy, giving rise to an implicit "first among equals" tendency: Paraphrasing Orwell, in modern bioethics all moral principles are equal but respect for autonomy is more equal than the others. By contrast, the "first among equals" spirit of autonomy in modern bioethics is not so evident in those few instances of surveys of global attitudes that are available. In the run-up to the Universal Declaration of Bioethics and Human Rights in 2005, the UNESCO International Bioethics Committee surveyed international organizations, national ethics committees, and other nongovernmental organizations to help identify what ethical principles should be included in the Universal Declaration. Although the numbers were small, "human dignity" (58) received more endorsements than "autonomy" (44).[11] While one shouldn't attribute more significance to this sort of result than it deserves, in my own experience, human dignity is a more central part of the bioethics conversation as one travels east and south than in the US and Western Europe. The use of terms like "dignity" and "autonomy" at such a high level of generality mostly flags different cultural perspectives, but they also shouldn't be waived aside. The Universal Declaration of Human Rights (1948) sets dignity as the primary moral reference point in unmistakable terms: "Whereas recognition of the inherent dignity and of the equal and inalienable rights of all members of the human family is the foundation of freedom, justice and peace in the world."[12]

Hostility to appeals to human dignity rather than autonomy, on the other hand, has appeared almost entirely in the American context. Reacting to a largely conservative, traditionalist approach exemplified by President George W. Bush's bioethics council, bioethicist Ruth Macklin depreciated the importance of the concept of dignity in bioethics, arguing that the equivalent conceptual work can be done by a principle of respect for autonomy. "Dignity is a useless concept in medical ethics," Macklin wrote, "and can be eliminated without any loss of content" in favor of autonomy or respect for autonomy.[13] Writing in support of Macklin a few years later, psychologist Stephen Pinker referred to "the stupidity of dignity."[14] A more moderate critique appears in Beauchamp and Childress's highly influential textbook, *Principles of Biomedical Ethics*, where they consider the view that human dignity is a property that conveys moral status, dismissing it as "an inexact notion that moral theory has done little to clarify."[15]

The assertion that respect for autonomy has become the first among equals in the universe of bioethics concepts that populate the rules-based order is not a criticism of Beauchamp and Childress. In their framework, the principle of respect for autonomy is nested in a set of principles in a nonlexically ordered list with a chapter devoted to the analysis of each one, respect for autonomy followed by nonmaleficence, beneficence, and justice. The early editions of *Principles* were criticized for presenting these four items as simply a checklist with little instruction about how to resolve conflicts among the principles in particular cases. They were to be "weighed and balanced" against each other. A secondary literature grew up in which the principles were to be "specified" according to the case, a strategy adopted by the authors.

There were substantive objections to what has come to be called principlism. As the Beauchamp and Childress approach gained traction with each edition, some moral philosophers argued that without a unified moral theory the principles continue to be at odds with each other such that that they are at best a useful checklist of moral considerations and at worst confusing in their conflicting claims.[16] Nonetheless, the utility of this list in the clinical setting amid sometimes crushing demands on a clinician's time and attention cannot be underestimated. The sociologist John Evans observes that in one clinical ethics handbook "the principles are so set that the book has a fourfold table with the principles printed on card stock, so that it can be put in your pocket when consulting in the hospital."[17] Or, as has happened in the intervening years, easily displayed on one PowerPoint slide. Again, this simplification can hardly be blamed on Beauchamp and Childress. The simple listing of the principles without analysis that Evans accurately describes is an institutional artifact that facilitates casual attention to the ordering of principles without appreciating the way their elements interact. In this way a richer conversation about the content of the principles is mostly a matter for humanities seminar rooms, not hospital hallways.

Nonetheless, the principles provided an accessible lingua franca for the nonexpert, or what John Evans calls "a common metric so as to allow communication in a pluralistic society." The principles also facilitated what I have called the migration of the principles from the research to the clinical setting, where they also achieved popularity as a mode of analysis, and not only at the bedside but in policy debates and in the regulatory system itself. "Part of the explanation for the dominance of principlism," Evans writes, "is that it spread from human experimentation to other debates in science and medicine. One reason for this expansion is the increasing number of issues with which the state or bureaucracies are involved."[18] "Principlism is the perfect system for these contexts because the principles are institutionalized," he argues, as "undebated ends that have a calculable quality and are described as derived from the 'common morality.' A new profession of 'bioethicist'—distinct from philosopher, theologian, and lawyer—arose that used these established and universal ends."[19] As well, "principlism has similarly taken on a life of its own, independent of the conditions that encouraged its early growth, although the continued appetite of the state and bureaucracies for bioethical decisions continues to encourage principlism directly. This independent life of principlism has been encouraged by the rise of bioethics as a profession."[20]

I would add one important qualification to Evans' observations: He writes mainly against a background of North American bioethics, where bedside clinical ethics consultations and hospital-based ethics committees are institutionalized to a degree not found elsewhere. In the US one can be certified as a "healthcare ethics consultant" (HEC-C), and the AMA has sanctioned ethics committees in health care institutions that "offer assistance in addressing ethical issues that arise in patient care and facilitate sound decision making that respects participants' values, concerns, and interests."[21] In such a regime a readily available, consensus-based listing of guiding principles is quite attractive for harried physicians facing complex matters of life and death, as is the availability of a "certified" ethics consultant. (These arrangements may, moreover, provide some protection from legal liability.) This extensive and "official" system of clinical ethics not only provides a kind of everyday concreteness that pervades and distinguishes North American bioethics; it also helps explain the prominence of a principle of respect for autonomy when actual cases and individual patients are involved. To be sure, efforts to introduce clinical ethics consultation in Europe have been made but, according to two Europeans engaged in that movement, "with a certain delay compared to North America."[22] One could theorize about the reasons for that delay, including the differing legal systems. Nonetheless, this a difference that matters within the culture and practices of bioethics.

Beyond Principlism

To be sure, Principlism has not gone unchallenged in the American academic discourse of bioethics. Libertarian H. Tristram Engelhardt Jr.'s bioethics theory prioritized autonomy over beneficence.[23] At the same time Edmund Pellegrino argued that physician beneficence was the moral core of medicine.[24] Bernard Gert, Charles Culver, and K. Danner Clouser objected to the inherent limitations of a conceptual framework that frequently generated incompatible results, as when the principle of respect for autonomy conflicts with the principle of beneficence.[25] Indeed, in the wake of the principles' influence a whole new literature sprang, intended to reconcile such conceptual conflicts by specifying the principles.[26] Arguing that no acceptable moral theory can eventuate in such conflicts, Gert and Culver proposed their own comprehensive approach to ethical problems based on ten moral rules.

Another approach sought to displace the role of principles. Stephen Toulmin and Al Jonsen urged the recovery of the tradition of reasoning known as casuistry, in which the crucial ethical elements of a case were compared to the details of others on a spectrum of greater and lesser moral acceptability.[27] For all their ingenious variety, in the end these other approaches did not fundamentally challenge a consent standard. Feminist bioethics did shift the focus in the direction of the larger context in which ethical problems in medicine arise. Often that context fails to take into account structural biases like the importance of gender, which often takes the form of undermining the ability of women to direct their own health care, including of course reproductive care.[28] That fairness has been marginalized in the modern health care system is also a central focus of the care perspective.[29] Despite the vigor of these critiques, the consent standard is not seriously questioned as a critical element in the analysis of bioethical problems, and principlism remains the dominant force in the world of bioethical theorizing, partly because in their successive editions, Beauchamp and Childress made strenuous efforts to incorporate key features of what could be seen as competing approaches. This is an implicit advantage of a looser-limbed conceptual framework than that urged by critics like Culver, Gert, and Clouser.

Besides the friendly cultural reception to the principles in the wake of the 1960s, the stars were aligned for autonomous consent (whether practicable or not) to be a dominant moral theme in the still-youthful American bioethics movement by the time the first edition of *Principles* was published in 1981. The first edition followed two decades of scandals about the rights of human research participants, culminating in the revelation of the US Public Health Service syphilis study (1932–72). Notably, some version of the autonomy principle had occurred independently both to Beauchamp and Childress and to the members of the first US government-chartered bioethics commission, the National Commission for the Protection of Human Subjects of Biomedical and Behavioral Research, in 1976. Although other sets of principles were up for discussion during the members' deliberations, in early 1977 Beauchamp was given the job of providing content to the "rough schema of three basic principles" that had been produced by the Commission's discussions. Beauchamp reports that he and Childress were already at work on the first edition of their textbook when he took up the role of staff philosopher at the Commission.[30] Indeed, in the Report, the Commission's respect for persons is analyzed in terms of autonomy, reflecting Beauchamp's dissatisfaction with a focus on personhood rather than on autonomy itself.

A personhood principle evokes a closer conceptual tie to dignity and to a more theological approach, inspired by conservative Christian philosopher Paul Ramsey. As assessed by a critic of the Beauchamp and Childress approach:

> In short, with *Principles of Biomedical Ethics*, autonomous and non-autonomous persons—so carefully held together by Nuremberg, Ramsey and the National Commission under the rubric of respect—are decoupled. Respect no longer pertains to the non-autonomous. Instead, their fortunes are determined by the principles

> of nonmaleficence and beneficence. Insofar as Beauchamp and Childress could describe these latter two principles in little more than utilitarian terms, their protections do not carry the moral security of respect.[31]

The key criticism here, whatever its merits, is that the prevailing form of the principles in Beauchamp and Childress's work is "little more than utilitarian" as it shifts the standards for the ethical treatment of the nonautonomous to nonmaleficence and beneficence rather than respectfulness. The decoupling of respect for the autonomous from the nonautonomous perceived by these authors concerning the *Principles of Biomedical Ethics* also signifies the diminished influence of theologians as US bioethics found its feet in the later 1970s. Earlier in the decade, the theologians, including especially Paul Ramsey, were central to the debate about the moral ends or goals of medicine. The reasons for the lesser role of theologians and the implications for the institutionalization of bioethics have been well-described by John Evans:

> While the public appreciated a debate about [moral] ends, the bioethics debate got the attention of the bureaucratic state, and soon there were government commissions and bioethical input into ethics policy at places like NIH, bureaucratic committees at research universities across the country and eventually hospitals. In these contexts, the consumers of bioethics, like the members of an IRB, do not want to debate ends like what a human is. They want generalizable, established and calculable ethical rules. Moreover, for policies that must be applied to all members of a society, these ends are best if they are portrayed as universally held by the public and not the opinion of some academics.[32]

The US Commissions Movement

So far, my narrative has reflected the leading American role for bioethics in the rules-based order, at least up to the turn of the century, but such a position isn't always pretty. With power comes responsibilities and vulnerabilities. Bioethics–American style appears to have succeeded beyond the dreams of its founders in the late 1960s and early 1970s. Or has it? In the words of historian Charles Rosenberg writing in 1999: "Bioethics not only questioned authority; it has in the past quarter-century helped constitute and legitimate it. As a condition of its acceptance, bioethics has taken up residence in the belly of the medical whale; although thinking of itself as still autonomous, the bioethical enterprise has developed a symbiotic relationship with this host organism. Bioethics is no longer (if it ever was) a free-floating, oppositional, and socially critical reform movement."[33] We have already seen the cooptation charge in a related context, from the standpoint of the "hard bioethics" movement's criticism of the role of the US in the creation and enforcement of global norms for human experiments and the FDA's ultimate separation from the Declaration's paragraph 29 in the year 2000 draft. From the domestic side, a parallel critical story could be told about the embracing of bioethics by the bureaucratic state, a story that has its best exemplar in the history of government bioethics commissions. LeRoy Walters traces the ancestry of modern ethics commissions to the British Royal Commission on Poor Laws (1832–34), which proposed reforms to the contemporary welfare system, based partly on empirical investigation. The Royal Commission became the model of a temporary advisory body with a limited or ad hoc government mandate intended to play a role in policymaking.[34]

Beginning in the 1970s with the National Commission, some governments' embrace of the commission model for bioethics was quite intense, especially in the West. A 2008 review of fifty bioethics commissions identified four characteristics: They worked on issues deemed important at a national level, the society supported the ethical inquiry to develop policy directions, they were "established or funded by a recognized national-level body," and "their members represented different disciplines, competencies and perspectives."[35] Sheila Jasanoff calls this "official bioethics," when lawyers, philosophers, and others participate in policymaking through government committees.[36] In one respect, this mode of bioethical influence avoids the question whether ethics is best left to experts or should simply reflect public sentiment or—put more positively—in theory governmental bioethics should combine ethics expertise with public input.

Thus, John Evans' mention of government commissions as the first in a list of entities that were institutionalized in the bureaucratic state is notable because of the unique role that commissions have had in legitimizing bioethics. One is hard-pressed to think of any other academic movement or scholarly field that has so benefitted from this recognition by the state at the very outset. The names of the first two US commissions track the widening agenda of bioethics beyond human participants research but now recognized by government. Evans points out that the "'National Commission' was 'for the Protection of Human Subjects of Biomedical and Behavioral Research.' The 'President's Commission' was 'for the Study of Ethical Problems in Medicine and Biomedical and Behavioral Research,'" with ethical problems in medicine a broader mission than research alone.[37]

The President's Commission (1978–83), which followed immediately upon the National Commission, also produced a series of highly consequential reports, creating guidelines (and in some cases inspiring state law) for informed consent, the definition of death, foregoing life-sustaining treatment, access to health care, and laboratory genetics, among others. Other presidential-level commissions were created in the Clinton, George W. Bush, and Obama administrations, also producing a variety of reports and recommendations, though perhaps none as influential as those of the first two commissions. There were also agency-level panels, like the Ethics Advisory Board (1978–79) of the Department of Health, Education, and Welfare (DHEW, now the Department of Health and Human Services), and the Human Fetal Tissue Advisory Board (1988) in the NIH. The presidential commissions' agendas tended to reflect the concerns of the administrations that created them. The Clinton-era formed an Advisory Committee on Human Radiation Experiments (1994–95).

During George W. Bush's presidency, the President's Council on Bioethics (2001–09) reflected conservative skepticism about superficial ethical consensus that undermined a "richer bioethics." Sometimes the commissions run into emerging controversies, as when the National Bioethics Advisory Commission (1996–2001) was assigned by President Clinton to help sort out human cloning and stem cell research or when President Obama's Presidential Commission for the Study of Bioethical Issues (2009–16) was asked to address revelations of human experiments on sexually transmitted diseases in Guatemala after World War II.[38] In the one case of a presidential bioethics commission that straddled two administrations of different parties, the Presidential Commission's report *Securing Access to Health Care* (1983) was reportedly caught in conflicting views about the role of government in providing health insurance as some Carter commissioners cycled off and were replaced by Reagan appointees.[39] A congressionally chartered Biomedical Ethical Advisory Commission (1998–90), to be composed of members of congress with an advisory board of "experts" and public members, produced no reports and was never fully established because of disagreements about potential congressional members' views on abortion.[40] In the US, the era of presidential bioethics commissions may have ended, at least unless and until the country's divisive politics are somewhat ameliorated. I have written about the place of consensus in bioethics at length elsewhere,[41] but it's not hard to notice the absence of a minimal degree of agreement about the functions of government and a willingness to accept that the "other side" has at least some good intentions, assumptions that more or less prevailed in previous years.

Commissions Abroad

In its early years one can perceive the influence of the American style of bioethics through the history of the field in Britain, a path that led to the creation of an independent advisory body called the Nuffield Council on Bioethics (NCOB). Medical ethics in Britain has a distinguished history. Thomas Percival's coinage of the term "medical ethics" in 1803 is appropriately celebrated as a milestone, but his stress was on the professional autonomy of physicians to police their errant colleagues. Sometimes called—without irony—"club regulation," in the wake of Percival it was up to the General Medical Council (GMC) to remove offenders from the registry. In the 1850s, the British Medical Association took the AMA's 1847 code of ethics as a model, itself based on Percival's ideas. The image of the physician as a virtuous gentleman to whom legal and ethical deference should be paid took longer to loosen there than in the US (reflected in the prominence of the professional practice standard), and when it did, British lawyers, and to a lesser extent religious leaders, played a more important role than philosophers, whose analytic orientation ignored or disparaged normative commentary. But as in the US, it was consent in the research setting that played the leading role in challenging orthodoxy. Nonetheless, as Duncan Wilson explains in his history of bioethics in Britain, until the early 1980s, "despite growing support for bioethics, the government only convened 'broad-based' inquiries to look into contentious new procedures such as IVF and gene therapy during the 1980s. . . . As before, they also had little say in the governance of clinical treatment. This offered a notable contrast to the United States, where hospital ethics committees that included bioethicists and 'community representatives' had the power to consider treatment and advise on individual cases."[42]

A key figure in the ultimate emergence of British bioethics in the American style, with its emphasis on the patient's voice, was academic lawyer Ian Kennedy, who was influenced by American bioethicists in his criticism of medical paternalism. Kennedy insisted that the medical profession itself could benefit from the involvement of sympathetic outsiders. These arguments did finally lead to the appointment in 1982 of philosopher Mary Warnock to chair the first British governmental commission that broke from control by the medical profession. The remit of the Warnock Commission, as it has come to be known, was specific but profound: the moral status of research and practice of in vitro fertilization (IVF). Their memorable and influential conclusion, that embryo experiments could be conducted up to fourteen days after fertilization (the so-called "primitive streak" that signals the beginning of the brain and central nervous system), was unsatisfactory to many reproductive biologists who found it too restrictive as well as to many philosophers and theologians who found it arbitrary. Nonetheless, following the model of American bioethics, the Warnock Commission was an exercise of government that rejected the notion that there were "moral experts" and that such questions needed to be discussed openly and in public precisely because they were in dispute.[43]

In the years after the Warnock Report, the trend toward involving nonphysicians in medical ethics deliberation gradually took hold in British medical education, including appointments of philosophers to medical faculty. Kennedy's call for a national ethics committee was realized in 1991, as described by Wilson: "After a series of conferences in 1990, the charitable Nuffield Foundation agreed to create an independent bioethics committee that included representatives from several professions. The resulting Nuffield Council on Bioethics embodied the belief that external oversight was vital to maintaining public confidence in biomedical research. Its establishment bolstered media support for outside involvement with medicine and science, leading the Guardian to claim that there was 'something of an ethics industry springing up.'"[44] The flippant (though not entirely inaccurate) remark about an "ethics industry" also appeared in US media in the 1990s.[45] In other respects, the public fascination with the ethics business followed the American pattern, as did the creation and original impetus for the Council itself, namely, public input on the ethics of research involving human beings. In the years since, the NCOB has reported on a wide variety of other topics, from mental illness to data science, just as the original National Commission in the US led to committees that ventured further into bioethics territory.

This writer will not dare to offend French honor by suggesting that the *Comite Consultatif National D'Ethique* (CCNE) was also inspired by the US National Commission and Presidential Commission, but the timing is notable. As in Britain, IVF was the impetus for the creation of the CCNE in 1983, which, like the NCOB, is a national advisory body but unlike the Council is government funded and has direct responsibility for revisions of bioethics law in France. Moreover, it is responsible for the revision of bioethics law in France, making it perhaps the most politically powerful of such body in the world. In terms of connection to government the three are quite different. The US commissions were, as noted, creatures of congress or presidential administrations. (In that respect they were exposed to allegations of political pressure, though I would say the relationships to presidential administrations were more philosophical than political.) The NCOB is, as noted, advisory only, while the CCNE president is appointed by the President of the Republic with members from dozens of institutions and, though funded by government, explicitly assures that it is independent of political pressure.

Their structural and legal status differences notwithstanding, the British and French organizations have proven to be especially adept at public education, at engagement, and at achieving a measure of consensus. But apparently some topics are thought better to be left to an ad hoc process rather than tying up the CCNE. Thus in 2018, President Emmanuel Macron announced the creation of a bioethics commission to review the country's policies on a wide range of subjects, including human reproduction, euthanasia, neuroscience, and artificial intelligence, with the CCNE as administrator of the process.

The German Ethics Council was created in 2007 by then-Chancellor Gerhard Schroder, replacing a previous National Ethics Council that had existed since 2001, that was less favorable toward preimplantation diagnosis (PGD).[46] In a pattern we have seen before, its remit is mainly in research ethics; indeed, in the German case the term ethics committee seems nearly always to represent research rather than clinical review. The president of the Bundestag (the parliament) appoints the members from a mix of disciplines who are compensated for their expenses, but they are to be totally independent of government in their deliberations.

Whether the French and German bioethics committees took any inspiration from the US experience is less important than that they represent a critical aspect of the place of bioethics in the rules-based order. They are part of a global network of national ethics committees (NECs) that are an irreducible part of the life sciences research and development system. The frequency of NECs might be surprising. A 2021 survey found that there are 124 NECs in one hundred countries, from Argentina to Zimbabwe: "The details of their missions as well as the formal, deliberative pathways through which they impact policy-making and societal debates vary. Some report directly to the government, a ministry, or contribute to legislative processes, whereas others provide non-binding counselling without pre-defined or guaranteed authority. One of the main functions of NECs is to facilitate public debate on controversial bioethical issues and to produce opinions and recommendations that can help inform the public and policymakers."[47]

Much of the data from this survey was drawn from a WHO database that was created following the 2010 Global Summit of National Bioethics Advisory Bodies, the 8th such summit, itself a remarkable statement of the globalization of "official bioethics" within the rules-based order. Higher-income countries predominate with 47 percent of the committees and, based on documents produced since 1975 (also the date of the first US National Commission report), 44 percent focus on research, though clinical and public health, and ethics are also addressed in NEC reports. From 1996 through the late 2010s these national bioethics commissions were part of a formal global network, a Global Summit of National Bioethics Commissions, that first met during the biennial International Association of Bioethics conference in San Francisco, organized by the National Bioethics Advisory Commission under President Bill Clinton.[48] Taken together, one can make out patterns that suggest a shared culture of international public bioethics deliberation. Most important, the very fact of these committees institutionalizes a consensus about certain values that reach beyond the laboratory into the clinic and the public domain, like respect for human dignity and personal autonomy. In this sense they establish a system of (often aspirational) values, especially within the vigorous, globalized, and lucrative medical research system.

Recently some of these national bodies' missions show a tendency to expand well beyond traditional bioethics topics like medical research ethics, especially in Western Europe. In 2021, the National Bioethics Commission in Greece, with eleven members appointed by the prime minister, changed its name to the National Commission for Bioethics and Technoethics. Its mission: "to examine the ethical, social and legal dimensions and consequences mostly of the fields of biology, biotechnology, medicine and genetics, and of new technologies, such as artificial intelligence, advanced algorithms and robotics."[49] A similar expansive tone for ethics commissions is evident in the language of the Danish National Center for Ethics, established in 2022, which observes the rapid advancement of technology, "increasingly demanding ethical considerations that span across research, data, health and biotechnology."[50]

Russian bioethics presents a fascinating contrast to these bioethics commission arrangements in liberal democratic states. From the Soviet era to the 2010's, a small group of Western-influenced bioethicists imported that orientation in the academic sector. Under the leadership of Boris Yudin, that work culminated in a project, "A Humanitarian Analysis of Biotechnological Projects for the Enhancement of the Human," that produced around 130 publications.[51] But according to Russian bioethicist Nataliya Shok, the importation of the secular Western approach didn't consider the Russian historical and cultural setting or moral guidance on medical technology. The post–Soviet political culture turned instead to the Russian Orthodox Church, which in 2000 created the Bioethics Commission of the Holy Synod, which is the de facto source of moral authority in the Russian Federation.[52]

Asilomar to Stem Cells

Governmental or national bioethics commissions are not the only means to achieve and provide consensus about norms. A necessary condition for their influence is often some underlying sense of urgency from the scientific community itself, especially when its autonomy is threatened by public controversy. In these circumstances nongovernmental or nonnational panels stemming from, and regarded as legitimate by, the scientific community may have at least as much leverage as entities that have the force of government behind them and, indeed, can help to prevent what the community would regard as political intrusion into the scientific process. Two examples in the US, separated by exactly thirty years but both created in reaction to well-publicized worries about laboratory science run amok, are the Asilomar Conference in 1975 and the US National Academy of Sciences (NAS) committee on human embryonic stem cell research in 2005. One might say that these panels, interwoven as informal actors within the rules-based order, produced the very softest of soft law in bioethics. They are not legally enforceable, are not presented as ethics codes, do not usually explicate underlying ethical or philosophical principles, and depend mainly on the extent to which the invisible college of science finds them to be acceptable guidance for whatever practical force they have.

Formally known as the International Congress on Recombinant DNA Molecules, what has come to be known among biologists simply as Asilomar took place at the Asilomar Conference Center in Pacific Grove, California. Experiments with DNA were novel and, to some, worrisome. Could the bacteria used to do studies of DNA escape from the laboratory? Among the platforms for these experiments were cancer-causing genes that could hold the key to new chemotherapies, as well as new biotechnologies for agriculture, and pretty much any field that could benefit from understanding genetics, but the possibility that they could represent threats to human health and the environment couldn't be dismissed out of hand. The colorful Cambridge Mayor Alfred E. Velucci, who had already sought to turn Harvard Yard into a parking lot, persuaded the City Council to enact a resolution asking Harvard and MIT to pause their research for three months while also establishing a local review board of scientists and citizens. It was, after all, only a few years after Michael Crichton had published his novel *The Andromeda Strain*. Shades of modernist worries from *Frankenstein* to the Wuhan lab, Velucci was not at all sure that "those people in white coats" couldn't let loose something catastrophic.[53]

In fact, led by Stanford's Paul Berg, the NAS had already recommended a voluntary moratorium on recombinant DNA experiments, especially those that could, in Berg's words, "change normally innocuous microbes into cancer-causing agents," or into antibiotic-resistant pathogens, or into potentially dangerous toxins. In the far smaller world of DNA research in 1974, that moratorium seems to have been universally, if somewhat grudgingly, accepted. Then, amid various scary scenarios in the press, Berg and several others obtained NAS and NIH funding to gather around 140 participants at Asilomar, a mix of scientists, lawyers, government officials, and journalists. Talk of restrictions on the science was not at all welcome at first, despite background worries that a failure to respond on the part of the scientific community would be followed by clumsy governmental intrusion into the scientific process. Finally, consensus was reached around certain practical measures that could set a standard for the work, such as physical containment for more risky experiments, or biological containment with bacteria modified so as not to be able to survive outside the lab. Those norms quickly become the basis of governmental biosafety guidelines and helped launch a thriving and remarkably productive biotechnology industry with none of the most feared consequences of modified microbes on the loose until, of course, allegations about the origins of COVID-19.[54]

Although the vast majority of the participants at Asilomar were from the US, there were a few from Australia, Canada, Great Britain, Israel, Japan, Poland, and even the Soviet Union.[55] Shobita Parthasarthy has described Asilomar as "a narrow affair" that excluded the Global South, triggering local resistance to what were seen as neocolonial practices such as genetically modified organisms (GMOs) that came with legal restrictions like patents and licenses.[56] But it took years for that response to crystallize. In the mid-1970s an American-led consensus on life sciences experiments, so long as it involved the major players and most prominent institutions, was in a strong position to dominate global practices. When the National Academies released its report *Guidelines for Human Embryonic Stem Cell Research* in 2005, those conditions still largely obtained. The Academies committee, which I cochaired, was created against a backdrop of uncertainty about acceptable laboratory practices for the derivation and study of stem cells from human embryos, a technology that had only been available since Jamie Thomson at the University of Wisconsin had developed a technique for isolating the cells in 1998. The George W. Bush administration had limited federal funding for the research of cells derived from embryos prior to August 4, 2001, but that left questions not only for many other details (particularly for nonfederal funding sources), like the conditions for research involving nonhuman animals and appropriate arrangements for obtaining and transferring embryos.

Despite the seemingly esoteric nature of this lab technique the topic signaled certain attitudes toward the moral status of the human embryo, thus the release of the report was accompanied by significant fanfare, including a press conference and a visit to Capitol Hill offices arranged by the Academies. In the now-quaint language of print news, the report was covered by the *New York Times* in an "above-the-fold" story on page one. The stem cell issue echoed themes of the Asilomar era, especially the publics' and the politicians' trust in science and scientists. Whether due to this exercise in the self-policing of science or not, there was no further legislative response, and, in any case, the development of embryonic-like stem cells (or "pluripotent" cells) from adult cells a couple of years later helped to forestall the political issue. The essential elements of the report did provide a template both for other countries' policies and for a newly founded International Society for Stem Cell Research (ISSCR) that soon published similar guidelines and has updated them in light of new developments in stem cell biology.

I conclude this chapter with one more example of the pervasive US influence in the standards for ethical conduct of basic biological research. Before genetic information is obtained from human biological samples, the donors are asked for their consent. There are different categories of consent depending on a spectrum of openness that ranges from publicly available data online to data that cannot be shared. These consent categories are not only used by scientific institutions in the US, but they are also represented in the organization of gene banks from Europe to China and defined in international and domestic law. Like so many other such consensus-based rules in the theorized invisible college of science, they originated in the US National Institutes of Health.[57]

These examples of soft law, or communal norms, apply to laboratory science rather than clinical ethics or research involving human participants. The next major controversy in lab bioethics took place a decade after the stem cell report, after various nonscience events in the early twenty-first century—among them wars in Iraq and Afghanistan, the 2008 financial crisis, and the rise of China as a peer competitor to the US—appeared to signal a relative decline of American influence in the rules-based order. Notably, an episode involving human genetic modification also raised questions about the continued influence of the US in the rules-based bioethics order and whether the scientific community would continue to speak with a single voice.

6
The Invisible College

During the Enlightenment, the Englishman Robert Boyle described himself as a member of an "invisible college" of correspondents, including Robert Hooke and Christopher Wren. They were part of an international group of "natural philosophers" (today called scientists and engineers) who wrote letters and traded books scribbled with marginalia, exchanges made possible through the emerging European postal system. In the twenty-first century, supercharged by the Internet, science is a global invisible college in which theories, methods, and data are instantly searchable and shareable. Pick up any major scientific journal and you'll note the scale and diversity of contributors, very often collaborators on the same important paper. Conferences in the largest fields draw tens of thousands of attendees where elbows rub, and partnerships are explored. Exchanges of laboratory personnel and research fellowship awards are not unusual. The result has been a distinct global culture of science, one that has coincided with the post–World War II rules-based order, in which scientists from distant parts of the globe often find they have more in common with each other than with their next-door neighbors.

In practice, like any human group, collegial aspirations often give way to less laudable motives and constraints. A website called

Invisible College invokes the spirit of Boyle et al., "the community of scientists who turn away from the visible metrics of success in status quo institutions (like publications, job titles, and awards), to focus on the invisible roots of science: curiosity, dialogue, and the subjective hunch."[1] But the constraints of visibility pull in another direction. Data are often closely held as proprietary, contributing to the "reproducibility crisis" in which many important experiments seem not to be replicable. Genetically modified animals are valuable forms of intellectual property that labs may be reluctant to share. When a "breakthrough" is announced it may come in the form of a press release rather than a peer-reviewed paper, seemingly designed to attract investors rather than scrutiny. Or it may come with much fanfare at an international meeting but not with the result the scientist might expect, as happened in Hong Kong in 2018.

Along Came Dr. He

Combined with a more or less overt disdain for politics, the global science subculture attracts a measure of suspicion from state authorities, or at least a watchful eye. Such a moment took place at the second world conference on the gene-modification technique known as CRISPR. The air was electric, and the room packed with journalists when a confident, youthful He Jiankui took the stage. A faculty member of China's Southern University of Science and Technology (SUStech), He had already posted a video announcing his use of CRISPR to modify the CCR5 gene in two IVF embryos that resulted in live-born female babies. The fraternal twins were named Lula and Nana. And as it turned out, a third baby's genes had also been modified.

The CCR5 gene is associated with susceptibility to a dominant strain of HIV, so He justified his experiment as a beneficent prevention of disease. But HIV-positive fathers don't pass on the virus, can have babies with in vitro fertilization so that they don't endanger the mother, and if the mother is infected the baby can be delivered by caesarean section to avoid exposure in the birth canal. The laboratory procedure was also faulty as the CCR5 gene was not completely edited. Moreover, the experiment could have unintentionally created errors in the babies' genomes, so-called off-target effects, exposing them to a higher risk of other diseases. And HIV is now treatable as a chronic rather than fatal illness, so there is not a powerful unmet medical need.[2] Other serious ethical failures quickly emerged, including the lack of adequate informed consent and a potentially fraudulent claim of ethics review.[3]

The international scientific community was taken off-guard by He's announcement and was, to put it mildly, highly critical. The reaction of a women's and reproductive health expert, Joyce Harper of University College, London, was typical: "Today's report of genome editing human embryos for resistance to HIV is premature, dangerous and irresponsible." Within China there was stark embarrassment, with 122 biomedical scientists condemning He's actions. "Directly jumping into human experiments can only be described as crazy," they declared.[4] A further consensus statement was produced by Chinese scholars in 2023, including among the signatories Xiaomei Zhai, a senior bioethicist in Beijing.[5] When I visited Beijing in December 2019, I heard similar condemnations from other bioethicists, who also noted that the whole episode had brought bioethics into prominence in China as never before.

In the immediate aftermath of the He case, some prominent scientists called for a moratorium on genome editing, reminiscent of the famous Asilomar moratorium on recombinant DNA research in 1975. Other distinguished researchers resisted that suggestion, apparently skeptical that a moratorium would be either enforceable (the global science world having become quite different since the 1970s), or that it would add significantly to the international opprobrium, including that of China's scientific community itself. Some may also have feared the precedent that a moratorium on a controversial scientific practice could set, perhaps threatening far more legitimate projects. The WHO created a registry for trials of human genome editing and formed an advisory group that recommended a global registry of preclinical studies.[6]

Some observers thought that Chinese officialdom would not be unhappy to take a different path, despite the risk of international opprobrium. In an incisive report journalist Jane Qiu noted that, according to several publicly availably documents, He's study that led to the embryo experiment was funded by three Chinese government grants, though it's not clear the agencies knew what exactly was being done with the money. "If the documents are correct," Qiu wrote, "they would suggest China is supporting research that the U.S. and other countries consider unethical, and raise doubts about the preliminary conclusion of a government investigation that He acted mostly on his own."[7] Writing in summer 2019, science journalist Jon Cohen, in a skeptical turn, said that so far "the institutional response in China has been feeble. The central government or other bodies have yet to conduct the transparent investigation many people had wanted. SUStech declared in its one statement on the matter that it wanted 'international experts for an independent committee to investigate this incident, and to release the results to the public.' But it does not appear that any such inquiry has occurred."[8] Others were more favorably disposed to the Chinese governmental response. Stanford's Hank Greely noted that despite inefficiencies in the Chinese regulatory bureaucracy, the Education Ministry had fairly rapidly called on research institutions to upgrade their management of gene-editing technology and tighter regulations were introduced to govern "high risk" technologies, as well as a national ethics committee that would report to the highest levels of leadership. Greely also warned against allowing racial bias to creep into assessments of China's handling of the affair.[9]

There were some other threads left hanging by the episode. Others, both in China and the United States, were aware of He's plans but were never sanctioned. Although the coin of the scientific realm in science is publication, considering the overwhelming consensus that the experiment was unethical, He's data has never been published, though based on a document posted on a Chinese pretrial registry, a geneticist has concluded that the editing did take place. Shielded from a curious international press, the three babies themselves are presumably well so far.

Although the production of living children surely sets He's experiment apart from other uses of gene editing, the international outrage it has attracted is something of a distraction from less boundary-breaking but nonetheless striking work in China. Taken together, these incidents suggest a pattern. A clinic in China has conducted gene editing in patients with esophageal cancer.[10] Another group created macaque monkeys with genes known to be active in human intelligence and speech.[11] These practices would have required a far longer approval process if proposed elsewhere, if they would even be approved at all. Even apart from embryos, the US CRISPR and other such genetic technologies fall under a tough approval process. A University of Pennsylvania cancer immunotherapy trial using CRISPR as an enhancer for the drug effects enrolled its first two patients in 2019, having jumped through various regulatory hoops over a couple of years.[12] But germline modification is largely banned in the most scientifically advanced countries. In the United States, the Food and Drug Administration imposes formidable barriers to clinical use of modified human embryos. There are also legislative obstacles to the use of federal funds for research with human embryos.[13]

In the aftermath, China created a series of regulations to prevent future cases like this, rules at least as formidable as those in the US, including amendments to the Civil and the Criminal Code. The National People's Congress's amendment to the Civil Code requires all human gene- and embryo-related research to comply with ethics and relevant administrative regulations, and an amendment was added to the Criminal Code that criminalizes serious illegal medical practices by unqualified personnel. Implanting gene-edited or cloned embryos risks a prison sentence up to three years. The National Health Committee must now approve gene-editing and assisted–reproductive technology experiments. And a new National Ethics Committee for Science and Technology is charged with guiding ethical governance of science and improving education about ethics in science.[14] Despite these measures, China watchers might note that ultimately the Communist Party controls state governance institutions.[15]

China's legal system found its way to these punitive measures in a more rapid course than would have been likely in the West. After the biophysicist lost his academic position and was subjected to house arrest, a court found that he had "deliberately violated national regulations," forged ethics documents, and "rashly applied gene-editing technology to human reproductive medicine," sentencing him to a three-year prison term and a $429,000 fine. Two of his Chinese collaborators were also fined and sent to prison for a time.[16] He was released in April 2022.[17] In the US some sort of fraudulent misconduct in the use of federal funds or a finding of criminal misconduct by a state medical licensing board might have ultimately resulted in similar civil and criminal penalties, although that is a matter of speculation. While the sanctions following the experiments were fairly immediate and intense as compared to a predictable US legal scenario, He's immediate postprison rehabilitation is also hard to imagine in a Western setting: Shortly after his release, He established a lab in Beijing with philanthropic support for the purpose of developing "affordable gene therapy."[18]

Genopolitics

This mixed response to the gene-edited babies affair is instructive. As compared to global norms that have been more formally adopted in other areas like armed conflict and trade, in esoteric fields of basic science the standards are often informal. To a great extent they depend on the professional culture of laboratory science itself. In that sense, adherence to the norms of biotechnology depends on a nation's willingness to abide by the spirit of the prevailing consensus of the international scientific community. The cochair of the WHO panel and the president of the US National Academy of Medicine noted that there is no international mechanism for halting these kinds of experiments.[19] Under these conditions, an important impediment to any inclination within the Chinese political class to go their own way in the ethics of biotechnology, unlike say challenges to the World Trade Organization and regional military alliances, is the culture of science in China itself. China, and the West in general, and the United States in particular, are deeply interwoven in the world of science, and especially the life sciences. However, there is a delicate balance between the advantage of a globalized extended community of science with a shared culture on one hand and national security interests and commercial interests (and, some would say, abject paranoia) on the other.

The various outputs of biotechnology and their promise for breakthroughs in agriculture, energy, and health care have stimulated a highly competitive global industry, one that did not exist in the 1970s but is now a key factor in economic growth and national prestige. Measured by the number of biotech firms, patent approvals, and biomedical treatment approvals, the most important national players are the US, South Korea, Germany, the United Kingdom, Japan, Mexico, New Zealand, and Belgium. In terms of research and development, the US is still the leader by far, followed by China, France, Switzerland, South Korea, Japan, Germany, and Denmark. But China began doubling its research and development (R&D) spending in 2008. As of 2024 China is ranked number one in global research output.[20] In biotechnology, one should add relatively substantial investments on the part of Singapore as well. Investment in biotechnology R&D is generally funneled through an academic-industrial research complex, although the details vary greatly from one country to another, such as how grants are disseminated and how incentives for intellectual property are managed. Military establishments in some of these countries also play a role, for example in funding defensive research on select agents and "dual-use" projects that could have both military and civilian applications.

Human beings have engaged in a traditional biotechnology called agriculture for at least twelve thousand years. The era of modern biotechnology began in 1971, when Paul Berg and colleagues at Stanford succeeded in using recombinant DNA technology to create novel strands of genes. Many have expressed concerns that modern biotechnology stands to create a radically new phase in the human experience. Novelists like Aldous Huxley in *Brave New World*, writing at the height of the eugenic era, speculated about the possibility that, whatever "genes" really are, they could be manipulated in a laboratory to produce different kinds, or castes, of individuals: Alphas, Betas, Deltas, and Epsilons developed in a Social Predestination room. These concerns were vastly sharpened in the nearly two decades between the publication of James Watson and Francis Crick's 1953 paper in which they reported the decoding of the DNA molecule and the advent of modern biotechnology two decades later. Questions about the implications of genetic control were among those that led to the field called bioethics in the late 1960s, just around the time that Berg and colleagues were approaching their milestone. That subject has been a touchstone of bioethics since the founding of the first bioethics research institutes, professorships, professional organizations, and journals in the late 1960s and early 1970s.

A US presidential bioethics commission was among the first to address the moral implications of the new biotechnology in a 1982 report called *Splicing Life*. The report identified themes that would reappear over the next decades and are now familiar, including the scientific community's self-regulation, the threat of "biohazards," the "Frankenstein factor," genetics and drug development, and various forms of direct manipulation of genetic material. As would be common for analyses produced in cooperation with the scientific community, the report downplayed vague worries about "playing God" and creating new life forms, although it allowed for worries about overdetermining a "designed" child's future, undermining traditional conventions about the biological family and equality of opportunity, and even changing the very meaning of being human.[21] The report warned that genetic modifications are no easy fix for social problems. And it hewed to a distinction between modifying an individual's DNA and that of their descendants, discouraging the latter out of fears of a modernized, government-managed eugenic program but otherwise not drawing a bright line. The potential benefits of gene splicing for the growth of human knowledge, for medical science, and for virtually every other domain in which genes are implicated were viewed as simply too great to give up with some permanent prohibition.[22] One might say that, even as the technology has progressed, the bioethical themes have appeared again and again, adjusted for specific cases and types of innovation. As a far more efficient system for modifying DNA than its predecessors, the gene-editing technology known as CRISPR has become especially salient.

Since 2017, the US National Academies[23] and the UK Nuffield Council[24] have published consensus reports on gene editing, both concluding that such modifications should only be done within strict regulatory frameworks and setting out several specific conditions including data on risks and benefits. Again, consistent with the history, somatic cell editing is generally viewed as manageable under current regulatory regimes for innovative medical interventions, with off-target effects a particular hazard. So-called "enhancements" are discouraged as unjustifiably risky and as poor investments of scarce research resources as compared to efforts to mitigate diseases. Germ line modifications, while not wholly ruled out if safety and efficacy can be established (a major obstacle considering the countless variables involved), are not seen as justifiable at this time.

The commercial and health possibilities implicit in modern biotechnology (in the largest sense the "bioeconomy"), particularly in precision and personalized medicine, have stimulated their own ethical issues. But these have mainly concerned privacy and data ownership, as vast quantities of patient information will increasingly be entered into databanks so that various datapoints (family history, genetics, response to medication, and other personal information) can be digested by machine-learning systems and translated into individualized treatment plans. Someday all patients might be "on protocol," a status historically reserved for participants in clinical trials, thus blurring the line between being a patient and being a research participant. With the important exception of well-defined, data-rich, and information-accessible diseases like cancer that have a certain range of treatment options and predictable progression, those systems are some way off. Thus, government investment in the basic science that can lead to such systems will continue to be critical in order to attract market players that see a plausible investment. And even in the case of cancer, the several hundred-thousand-dollar per-patient annual price tag for the promising new immunotherapies in which CRISPR can be utilized is giving pause to both the pharmaceutical industry and private and public insurers.

Follow the Money

Beginning around the time of the He Jiankui episode, efforts to build China's life sciences capacity became the tip of a legal spear in responses by US law enforcement against suspected unlawful activity. The Thousand Talents program, established in 2008, aims to establish China as the world's technology leader by creating financial and other incentives for talented scientists to remain in the country, to return from a period abroad, or for non-Chinese scientists to establish labs there. He himself was a beneficiary of a related effort called the Peacock Program, with $6 million to launch a startup to sequence single DNA molecules.[25] Thousand Talents has been implicated in the arrest of several scientists in the US on charges of espionage.[26]

As this unfolded, a former director of the US National Institutes of Health, Elias Zerhouni, warned that years of productive collaboration between the two countries is under threat if these kinds of sanctions are indiscriminately applied. Writing in the journal *Science*, Zerhouni argued that "the United States should not risk losing critical intellectual assets such as productive foreign-born scientists and engineers to global competitors to serve short-term security concerns at the expense of long-term national interests."[27] Unlike the Soviet Union during the Cold War, the life sciences community in the US is deeply engaged with scientists in China who are on temporary academic assignments in American labs and with many Chinese nationals who have immigrated from China. How to separate legitimate scientific work from intellectual property issues and worries about espionage amid the implications of simmering racism will be a continuing challenge. In 2019, more than a dozen scientists working at US universities were referred to federal officials due to concerns about foreign influence and dozens of institutions were contacted about these suspicions, stoking anxiety among Chinese scientists in the US.[28]

Apart from security and intellectual property concerns, in the months after the He Jiankui scandal there were signs that the loosening of assumptions about moral norms regarding genome modification might already have occurred. One might have thought that the prison term for He Jiankui would have discouraged imitators, and perhaps it did so in China. However, a Russian scientist said that he would use a superior embryo-editing technique to address CCR5 and AIDS susceptibility.[29] Denis Rebrikov of the Pirogov Russian National Research Medical University then switched his plans to addressing a gene implicated in hereditary deafness, but was met with a storm of criticism, including a high-level meeting at the Russian Academy of Sciences philosophy institute that seems to have included one of

Putin's daughters, a pediatric endocrinologist. Once again, it was the leadership of the Russian scientific community in insisting on the application of their extant regulations that chastened Rebrikov, regardless of Moscow's official views on matters of global norms.[30]

Finally, as we have seen, despite the speculation that China might go its own way on germline genetics, officials there did not propose different rules as the basis for international standardization in research involving human gene editing. Had it done so, or if it does so in the future, it could assert itself as a leader in thinking about the ethics of genome modification as part of its role in defining a new international order. Such a competition has been going on in the artificial intelligence (AI) sector for some time. A 2018 white paper on AI by China's Standards Administration sets rules on AI safety, ethics, and privacy: "Establishing policies, laws, and a standardized environment in which AI technologies benefit society and protect the public interest are important prerequisites for the continuous and healthy development of AI technology."[31] In July 2019, in response to a presidential executive order, the US National Institute of Standards and Technology issued a draft plan communicating that "AI standards that articulate requirements, specifications, guidelines, or characteristics can help to ensure that AI technologies and systems meet critical objectives for functionality, interoperability, and trustworthiness, and perform accurately, reliably and safely."[32] The US has stepped up both civil and criminal legal actions targeted at China's AI ambitions, alleging theft of trade secrets and hacking into personal data held by consumer credit agencies, hotel chains, and Microsoft servers.[33]

Can the competition for an advantage in AI be distinguished from genetics? This comparison illustrates the possibility that global rules can be sustained in some sectors but not in others. Unlike the implications of advances of various forms of AI without a global standards regime, the risks of systematic, lab-based human germline modification are at best highly speculative. Skeptics about the weight of moral norms against deliberately modifying heritable traits might also note that such alterations might not only be imperceptible but would pale beside natural de novo variations that take place in each individual genome and are passed down through generations. Unlike humans, rapidly propagating species like mosquitoes are not only vastly easier to modify in large numbers, they are also of far greater interest for population health. In studies that are going on now, a form of gene editing known as gene drive could deliberately reduce insect populations that are the vectors of diseases like malaria and yellow fever. Similar modifications could reduce populations of rodent pests in agricultural areas.

Two areas of biotechnology that do demand a vibrant international normative and institutional order are those of laboratory security and biological weapons. Managing the results of experiments in microbes and in animals through safe laboratory practices, regimens to monitor and control "dual-use" research in select agents, and cooperation to prevent the creation of new and virulent threats as weapons of terror should continue to motivate nations to maintain a basic normative regime for biotechnology. Whether such a regime can coexist with one in which values concerning human reproductive materials are viewed quite differently is at the moment both an open question and, potentially, a stress test for the normative scope of the international order as it applies to reproductive genetics. Take the case of DNA profiling, where it's far from clear that the cultural norms of the global invisible college can resist the operations of a surveillance state: In 2022, a genetics paper published in the prestigious journal *Public Library of Science (PLoS)* raised suspicions about blood samples (biometric data that could be useful to security forces) collected from hundreds of people who are among oppressed populations in Tibet. Not convinced that proper review and consent procedures had been followed, the journal editors retracted the paper in 2023.[34]

Globalization and Its Limits

I began with the case of He Jiankui because it illustrates how the global cultural community of science operated in the face of a bioethics challenge, largely speaking with one voice, and ultimately joined by the political system that did not choose to separate itself from the scientific community's consensus. So far, then, there's no occasion here for the emergence of geopolitical tensions. Another episode in which the modern, globalized culture of science and the national political system spoke with one voice took place in 2005. Seoul University scientist Woo Suk Hwang had falsified research claiming that he and his team had produced eleven embryonic stem cell lines containing nuclear DNA from body cells. It turned out that, not only were the cells obtained from two junior members of his lab (a violation of the power inequalities in that setting), but two of the photos in the *Science* paper were duplications, leading Hwang to confess that the evidence was fabricated.[35] By then Hwang had become a national hero in South Korea, having had a stamp created in his honor and given free lifetime travel on the national airline. The outrage was, again, global, swift, sure, and thoroughly consistent. In this instance, however, there was no suggestion among any observers that national prestige could endorse outright fraud.

The damage to South Korean national honor was driven home to me during Christmas week in 2005. I was asked to meet with a member of the South Korean science academy and an official of the science ministry at the National Academies building on Constitution Avenue in Washington, apparently because I had cochaired the Academies' recent report on embryonic stem cell research, and because I was conveniently local. Despite my very modest status (I am of course not even a scientist), I sat in a largely empty building in the Members Library for about an hour solely to receive the heartfelt apologies of the South Korean colleagues for the Hwang incident.

Many more such examples of the way that a certain consensus about practice standards are an element of the globalized culture of laboratory science. Of course, the Hwang case was quite different from the He Jiankui case; fraud is a far more familiar and settled ethical violation even than the manipulation of embryonic DNA without sound justification. The fact that both of these cases took place in Asia, where by the turn of the century highly sophisticated science cultures had established themselves, stands in contrast to the Asilomar conference in 1975, generally thought to be a landmark of scientists' self-policing, where virtually all of the attendees were from the West.

Crucially, Asilomar was also an expression of the scientific community's concern to keep the US congress at bay, to allow science to be governed by scientists. This is a familiar pattern in post–World War II science. A few years after the publication of the Nuremberg Code following the Doctors' Trial in 1947, the World Medical Association began a decade-long process of developing its own Declaration of Helsinki. The Code, though influenced by input from the trial's medical experts, was written by judges. The WMA wanted to make sure that medical doctors were in control of their own ethical standards, hence the Declaration. Both Asilomar and the Declaration reflect the view that political authorities needed to be reassured that the professionals wouldn't embarrass them by allowing sensitive matters like human experiments to get out of hand and risk public embarrassment.

The larger point is that the voice of a rules-based order can, and largely has, functioned in the life sciences—though surely not in all cases—along with the internationalization of science. (Journal editors and funding sources can play an important enforcement role when they insist on standards like transparency and consent; unfortunately published disclosure policies concerning conflicts of interest are often not honored.)[36] So long, that is, as that order is perceived by authorities to be consistent with the state's political interests. It is when those interests collide with the globalized culture of science, as has been the case during the coronavirus pandemic, that the friction between bioethics and geopolitics becomes most apparent.

This conflict is not new, though the most familiar cases are so familiar and so obvious that they are easily overlooked. Earlier I alluded to the history of human experiments by national security agencies during the Cold War. Even in cases where violations of medical ethics have been tacitly acknowledged at the time, political interests have often prevailed. The grim litany of such cases are well-documented elsewhere.[37] Here I aim primarily to call attention only to the fact of bioethical-geopolitical tension as rendered visible by these episodes. They help us perceive bioethics as framed by larger interests of political power and nation-states' interest in superiority over peer competitors and, in the extreme, existential survival.

International conventions about science and medicine in armed conflict and for national defense, such as limitations on the stockpiling and use of biological and chemical weapons, are mainly prudential. They set forth rules concerning what are generally considered to be exceptionally inhumane (and largely inefficient) weapons of war, such that each potential adversary should be able to set aside certain worries about sources of suffering and death of its own forces and of noncombatants. But those conventions are also bioethically relevant precisely because they identify implicitly moral limits. It goes without saying that these rules have been violated in spirit if not in letter, but they are still expressions of a certain rules-based order. The same is true of the LOAC that incorporate those conventions and others.

Admitting the limitations of explicit international conventions, their absence in the context of a global pandemic for which the world was, it turns out, poorly prepared, has established an even more glaring example of the way geopolitics creates the conditions for bioethics in practice. Unlike the exotic case of the laws of war, a conversation usually left for experts, the pandemic forced itself into the awareness of just about everyone. In doing so it also laid bare the geopolitical conditioning of certain bioethical standards and concepts. Domestically, the provision of ventilator support for COVID-related complications in the intensive care unit (ICU) was seen by some observers as inequitable due to risk factors that predominated in Black patients, like asthma and diabetes, risk factors which were in turn attributable to generations of deprivation from the effects of structural racism.[38] Equity concerns surfaced, as well, regarding access to vaccinations when they became available.[39] Sometimes, as in this case, the public health value of equity aligns with a less lyrical value in public health: utility. For, those who were more likely to be treated inequitably by reason of underlying medical and social conditions were also vulnerable to the dangerous complications of COVID-19, with all its attendant threats to others in the community. These arguments were, of course running side-by-side with the Black Lives Matter protests and headfirst into the mask wars and conspiracy theories.

Not only was the value of equity controversial ground but so was the value of utility. Utility in public health means the reduction of morbidity and mortality. Recommendations with those ends in view are to be based on the best available science. The trouble was that the science concerning the behavior of this new coronavirus was (and to a great extent remains) unsettled. In turn, the utility of expert recommendations was (and, again, to some degree remains) somewhat limited. If there is one lesson to be learned by public health agencies from the COVID-19 pandemic experience it is, as Jonathan Marks argues, that they must "explain the evolving nature of our knowledge and understanding and the evolving nature of the threats to both individual and public health." The failure of what Marks calls "consistency at any given moment and constancy over time"[40] in public health communications exacerbated other problems, including a longstanding distrust of scientific expertise, the adverse effects of lockdowns on local economies, and both institutional and political obstacles to data-gathering about the presence of the virus in the community.

Equity and utility make sense at the level of global public health as well, but in that context the story was a comparative one. There the limits of political authorities' tolerance for the values and preferences of the scientific community expressed itself in a variety of ways, mixing in a devil's brew of complexity with the effects of the virus itself. Welfare states in Eastern and Western Europe largely rallied their populations. Britain and the US did not. National wealth was no predictor of the scale of the epidemic, as Italy and Spain suffered grievously in spring 2020 but not Pakistan or Vietnam. Some small rich Asian countries fared well at first, but so did some small poor Asian countries. Sweden chose one sort of trade-off: lives versus keeping the economy open; Denmark chose another. Which was the "winning" strategy cannot be determined by the data alone. Gross geopolitics didn't seem to make such difference in predicting winners and losers.

7
A New Order

The COVID-19 pandemic, wars in Ukraine and the Middle East, and widespread disaffection with liberal democracy have accelerated trends undermining the post–World War II rules-based order. The instability of the imperial system finally led to the array of institutions and norms with which we have become familiar and now appears to be in jeopardy. Contrary to the most optimistic forecasts in the 1990s, promulgated by so-called neoliberals, more efficient markets and open trade did not draw China or the new Russian Federation into liberal democracy. Nor did the authorities governing the rules-based order step in to correct the excesses of global capitalism. The result has been an insidious feedback loop: The forces stimulated by deglobalization have produced a degree of institutional decline and neo-imperialism, which have in turn intensified inward-facing and isolationist tendencies around the world.

For well or ill, there can be no turning back to the brief era of US hegemony from the 1990s to the early 2000s, when one noted writer perceived a triumph of liberal democracy as the "end of history." As, a paradigmatic creation of the rules-based order, the outstanding question is whether adjustments need to be made in the standard ways of analyzing bioethical problems. Academic bioethics, especially in the

US, has produced forms of analysis that rode on waves that included the emergence of controlled clinical trials, more powerful laboratory technologies, and an explosion of knowledge and investment in basic science.

Global Bioethics?

Under the rules-based order diverse institutions and norms have created a varied system for global bioethics governance; bioethics itself as a successor to traditional medical ethics needs to be understood as a creature of that system. To say that American bioethics has been a dominant force in the field is not to say that no globalist alternative has been offered. In the ambitious 2013 *Handbook of Global Bioethics*, with an international array of dozens of authors, the editors, Henk A.M.J. ten Have and Bert Gordijn, set out their vision: "Since the turn of the millennium, bioethics has entered a new, more expansive stage. Due to international cooperation, new information technologies, and transnational economic activities, bioethical issues have also been globalized," mainly in relation to the inability of low-income countries to afford the modern medical care available to others. Granting the Western origins of bioethics, they argued that problems like poverty and climate change require both a global perspective and the recognition of diverse cultural perspectives. Developing countries especially have wanted "a common framework of ethical principles," but the editors also insisted on respect for local values. Beauchamp and Childress's principles, they asserted, can be complemented with non-Western values like solidarity, social responsibility, and benefit sharing. In this way "a global geography of moral values," in association with "universal human rights discourse enables humanity itself to be regarded as a moral community. It implies that citizens of high-income countries can no longer be indifferent to obscure clinical research practices or organ trade in low-income countries, since the same moral values and standards apply within the global community, although the application of principles is always modified according to local circumstances and local communities."[1]

After a tumultuous decade, the editors' vision of a global moral community of bioethics seems sanguine indeed. Nor, despite their efforts, were they able to reconcile some fundamental tensions between Western and non-Western values. Informed consent, for example, is asserted to be universally required, but "in different cultures informed consent takes shape in different ways."[2] Some African countries take a communitarian approach such that the "community leader takes the lead in decision making."[3] However, on some versions of such "lead taking" the shape of informed consent could be distorted to the degree that it is at odds with the status of women or ethnic minorities and not merely a different application of the same values.

Although ten Have and Gordijn struggle mightily and admirably to make their case, their account of a truly globalized bioethics, like the invisible college of science, is mainly aspirational. In reality, the Western-dominated bioethics framework is deeply and formally embedded in the international system of medical science R&D. Try to imagine, for example, the pharmaceutical research and development system with an informed consent standard that permits consent by a powerful local personage, usually male, rather than the competent prospective participant. In a diminished or fractured international order, some of the now familiar assumptions underlying bioethics as a functional value system for medicine and the life sciences would be at hazard, most especially including universal human equality and the rights to universal standards of safety, professional responsibility, and informed consent that it implies. What would be the consequences of a diminished rules-based order regarding those assumptions?

One casualty could be voluntary, consensus-driven documents that are intended to set global norms regarding specific scientific practices, the "rules of the road." An example is the response to the advent of embryonic stem cell research, made possible by the isolation of human embryonic stem cells in 1998. In this instance the technical breakthrough stimulated a political debate about the moral status of the human embryo, because while obtaining the cells the embryo is destroyed. When President George W. Bush established a compromise that permitted research on already available "lines" of these cells in 2001, the life sciences community was uncertain about what other specific procedures were appropriate, such as whether it was permissible to introduce human embryonic stem cells into laboratory animals.[4] After a National Academy of Sciences committee that I cochaired issued the first set of guidelines for this research, there was a felt need to establish international guidance. The ISSCR was created for that purpose. The ISSCR guidelines have been updated several times to accommodate new technical possibilities.[5] They are not in themselves tied to enforcement mechanisms or sanctions if they are violated, an inherent weakness of the invisible college of science, but at least legitimate sources of funding, investors, regulators, and venues for publication know where to look to determine the consensus-based boundaries.

Also at risk in a diminished rules-based order would be the plausibility of "anticipatory governance," an aspirational notion frequently referenced in bioethics, especially at the international level. These efforts differ from those that are introduced after the fact, as in the case of the ISSCR guidelines. As defined by David Guston, anticipatory governance is "a broad-based capacity extended through society that can act on a variety of inputs to manage emerging knowledge-based technologies while such management is still possible."[6] Among those inputs is the empirical study of societal preferences for the management of emerging technologies, which could take the form of focus groups or surveys. Though as Guston notes, anticipatory governance could be dismissed as a public relations campaign by self-interested parties to avoid hostile public reactions, the management goal is primary and laudable. In fields like gene editing, the rapid pace of transformations in medical science and technology have given rise to attempts to anticipate the implications of disruptive innovations and to plan accordingly. (From 2020 to 2023 I participated in one such ambitious project on gene editing that was conducted by Baylor College of Medicine with NIH funding.)[7] However, any attempt to get ahead of the ethical challenges of emerging science and technology presupposes a rules-based order that can implement either new regulatory structures or at least introduce novel consensus-based norms with some reliability. Another route is the adoption of moratoria on the pursuit of certain technologies, from recombinant DNA to genome editing to artificial intelligence-based systems. But again, proposals for moratoria would be empty without an order to which to appeal and within which new norms can operate.

US Security and Insecurity

Regardless of what form it takes, self-governance of science has its limits; one of them is a nation-state's security interests. For seventy-five years the conditions of US national security policy established by presidential administrations have had ripple effects through the scientific activity of each era. A least three times since World War II, US national security policy has been significantly revised, once to deal with Communism, then with terrorism, and most recently with the threat to the rules-based order. President Truman's policy expressed a commitment to strength to counter the threat of Soviet expansionism to which every subsequent Cold War–era administration subscribed. President George W. Bush's 2002 policy statement for the GWOT addressed a new problem: "America is now threatened less by conquering states than we are by failing ones. We are menaced less by fleets and armies than by catastrophic technologies in the hands of the embittered few. We must defeat these threats to our Nation, allies, and friends."[8] Exactly twenty years after the US national security policy that emphasized the challenge of failed states, in 2022 the official policy struck a somewhat different tone: that America "will partner with any nation that shares our basic belief that the rules-based order must remain the foundation for global peace and prosperity."[9] The 2002 policy has the feel of a statement issued by a status quo power nation seeking a strategic readjustment of its security posture. By contrast, the 2022 policy language has the feel of a global power on the backfoot, insisting on preserving normative conditions that seemed safely assumed not long before.

Dissimilar as the circumstances are in geopolitical terms, in the adjustments to the Cold War, the GWOT, and the defense of the rules-based order, there are striking similarities as well. Repeatedly, national security institutions were remodeled and given new powers, the military establishment's mission was reviewed and altered, and science was identified as playing a key role in providing an advantage to the US and its allies. The net result in each case was a significant expansion of state power by both major political parties, despite the usual anti-government rhetoric. But expanded state power does not automatically translate into societal acquiescence, especially in matters affecting the public health. This is not only a pandemic-era phenomenon: In the early 2000s clinical trials of smallpox vaccine were supposed to lead to an immunization program for emergency responders. According to the Bush administration's plans, by July 2003, five hundred thousand military personnel and an equal number of civilians were to have been immunized, but instead, reflecting a distinct lack of enthusiasm for the effort, only 10 percent of that number had received the vaccine. Nearly fifty thousand doses of vaccine had to be destroyed because the contents of opened vials had expired.[10]

This result was in some ways reminiscent of the public reaction to another emergency response plan during the Cold War. Initial enthusiasm for a fallout shelter program during the Kennedy administration was represented by a several hundred-million-dollar civil defense appropriation request from Congress. Despite an intense campaign to convince Americans that their survival could depend on a shelter in their homes, the effort quickly faltered amid confusion about the true benefits of a shelter in the event of thermonuclear war. The federal government permitted the initiative to fade away. By contrast, current efforts to restrict access to laboratories, materials, and scientific information, and to manage the visa status of laboratory researchers, have been met with skepticism and even alarm on the part of the scientific establishment. In December 2002, the National Academies expressed concern that constraints on foreign visitors have had "serious unintended consequences for American science, engineering and medicine."[11]

Twenty years later, as worries grew about China's exploitation of the openness of Western science, the Justice Department's China Initiative investigated scientists of Chinese descent or who had funding from Chinese entities. A chill ran through major research universities when their leaders received visitors from the FBI, warning them to monitor Chinese scholars and students. As many as one thousand Chinese researchers left the US; others had visas revoked. And of the dozens of cases filed only a few resulted in findings of guilt, mainly having to do with improper filing of tax forms.[12] The China Initiative formally ended in 2022, but for the US the underlying problem of technology transfer with China remains, especially with the ascent of President Xi Jinping, who has departed from his recent predecessors in emphasizing security over growth. According to the Center for Strategic and International Studies, "The incidents of Chinese espionage far outnumber those of any other country, including Russia. . . . Chinese espionage also created immeasurable damage to national security with the theft of weapons technology, including nuclear weapons test data. In the last few years, China has added the theft of massive amounts of personal information (PII), political coercion, and influence operations to its espionage activities."[13] If this assessment is accurate, it would be naïve to assume that the life sciences and biotechnologies do not fall within the zone of security competition. But those familiar challenges may assume a very different form if open source models like those created by China's DeepSeek prove to require both far fewer chips and less centralized policy control over AI. Whether directly or indirectly, bioethics is not immune to great power rivalries or to the disruption of the familiar rules-based order.

Vaccine Nationalism

Despite a pre-COVID-19 consensus that Western countries were well prepared for a global pandemic, especially the US, even wealthy and apparently well-organized systems both under-performed and failed to contribute a fair share to the international structures in place.[14] In 2020, with substantial at-risk investment by certain governments, vaccines were tested and produced in an astonishingly short time; industrial capitalism, the pharmaceutical industry, and academia together did well doing what the incentives aligned them to do. Allowing for some differences in timing that were critically important in spring 2021, by that fall just about every northern hemisphere country in "the West" had more vaccine than the global south. However, certain requirements for a successful distribution effort received less attention, including consistent production quality[15] and effective delivery systems even within the US.[16] And when vaccines are plentiful and production quality assured, infrastructure limitations that undermine distribution in many low-income countries remain prevalent, an example of the "last mile" problem, one familiar to anyone trying to deliver a package under less-than-ideal circumstances. Referring to the worldwide initiative for equitable access to COVID-19 vaccines, the WHO director-general said in 2021, "The problem is not getting vaccines out of COVAX. The problem is getting them in."[17]

A different sort of problem, and one that requires more attention in bioethics, was the emergence during the pandemic of "vaccine nationalism," in which vaccine-producing countries vied for control, and nonproducers were dependent on markets and political leveraging by the producers. Granting that public health is inherently political in the sense that it always involves managing or at least tracking the interactions of bodies, both human and nonhuman, the pandemic vividly demonstrated the consequences of a diminished rules-based order. Some cases of what came to be called vaccine nationalism were more gross than others. The case of Hungary is especially interesting because the Orban government, though a member of the European Union, had already charted its own course on matters like the 2015 immigration crisis and a self-described "illiberal" ideology. Criticized by the European Commission for deviating from "rule of law" standards, Hungarian authorities saw an opportunity to achieve mass vaccination more quickly than the rest of Europe. Thus, Hungary looked for alternative sources of vaccines apart from the EU joint effort. When a deal was struck for Russia's Sputnik V, the vaccine produced by the Gamaleya Research Institute of Epidemiology and Microbiology, one writer called the vaccine a "weapon for global security,"[18] and another "a tool of soft power for Russia."[19] Vaccine nationalism expressed itself in other countries as well. In Slovakia, Prime Minister Igor Matovič's nationwide testing program did not achieve the expected success in the fight against COVID-19 (in what was already one of the countries with the highest incidence in the third wave, along with the Czech Republic and Hungary), but the crisis was worsened by the government's decision to override the objections of its coalition partner and secretly purchase two million doses of the Russian Sputnik V vaccine that had not yet been approved by EU regulators. Following a wave of public criticism from members of the governing coalition for having offended the country's European partners, and with doubts being raised in EU capitals about Slovakia's "pro-Western orientation," Matovič had to resign at the end of March 2021, leading to further political instability at a crucial juncture in the pandemic.[20]

There were irregularities in other parts of the world as well. In Argentina, the government reached an agreement with Russia to receive Sputnik V. But the Russian government failed to deliver vaccines, which was one reason that only 11 percent of Argentina's population had been vaccinated by May 2021.[21] Nor was the US immune from execrable behavior. In 2024, it was reported that the Pentagon had engaged in a disinformation campaign "to sow doubt about the safety and efficacy of vaccines and other life-saving aid that was being supplied by China," a report that was not denied by US officials.[22] Putting it mildly, the science diplomacy aspect of the rules-based order did not function up to its ideals under the pressure of time in response to a highly transmissible, but poorly understood, upper respiratory virus. At the most basic level, countries couldn't even standardize reporting data, like health outcomes by age and sex.[23]

Despite the dispiriting symptoms of vaccine nationalism that almost certainly resulted in millions of unnecessary illnesses and deaths, it's worth pointing out that in many instances valid scientific processes showed resilience during the pandemic. Among the regulatory processes that continued to function were those of international bodies like the World Health Organization and the European Medicines Agency, and national regulators like the FDA and the South African Health Products Regulatory Authority. Approved vaccines continued to require the core elements of ethical trials, prior review, and informed consent. In 2021, the director of China's Center for Disease Control and Prevention admitted that their vaccines "do not have very high protection rates," and gave due recognition to the mRNA-based vaccines.[24]

Vaccine nationalism doesn't always become expressed through overt competition; it can also appear as a form of generosity. The geopolitical leverage provided by vaccination campaigns manifested itself quite differently in one episode during the Cold War when the Sabin vaccine was adopted by Soviet and Eastern European countries experiencing multiple polio epidemics. Renamed the

Sabin-Chumakov vaccine for Sabin's Russian partner, in 1959 Hungary was the first country to implement a mass vaccination campaign with the live virus formulation soaked into a sugar cube.[25] In 1961, the East German government benevolently offered Soviet-produced supplies to its Western counterpart, an unsubtle propaganda message about the supposed superiority of socialist science and public health organization. Ultimately the West German federal health agency decided that the oral polio vaccine was not yet supported by sufficient data from clinical trials, taking Chancellor Konrad Adenauer's government off the hook.[26] The rules-based order often works in mysterious and wondrous ways.

One group of bioethicists argued that neither nationalism nor "extreme cosmopolitanism," in which governments claim no special access for their own people, is ethically justifiable. They suggested instead a "fair priority for residents" framework "in which governments can retain COVID-19 vaccine doses for their residents only to the extent that they are needed to maintain a noncrisis level of mortality while they are implementing reasonable public health interventions."[27] While we may agree that the producer states had the right to consider first the needs of their own populations, in many cases an aura of excessive self-regard pervaded the response.

The Russia Test

A look back at the Russian Federation's COVID-19 response provides a useful "stress test" for global public health in a diminished rules-based order. As the pandemic response unfolded Russian authorities not only failed to meet customary safety, efficacy, or production requirements (the process did not include phase III trials with a diverse group of volunteers) but sought to distract attention from their system's incapacity. Acknowledging that all producing countries were rushing through the process due to the public health emergency, the Carnegie Endowment for International Peace nonetheless singled out the Russian project as exceptionally geopolitically motivated: "Instead of tackling such [regulatory] challenges head on, Sputnik V's backers and the Russian state-controlled propaganda apparatus have repeatedly tried to tarnish the reputation of Western vaccines. For example, when Pfizer released data admitting decreasing efficacy of its COVID-19 vaccine over time as part of the U.S. Food and Drug Administration's regulatory review of its booster shot, Sputnik V's Twitter account made a specious comparison to erectile dysfunction medication."[28]

According to Russian bioethicist Nataliya Shok, the larger context of the Russian approach to its vaccine rollout is President Putin's "use of public health policy prioritizing international influence and security." As Shok notes, Russia did not follow international protocols for clinical trials; ignored, and still does not have, WHO approval; and therefore cannot use international distribution networks; including WHO's COVAX system. But from the Russian state's standpoint, the rollout was a success. The message of Sputnik V was made clear in its very name, that Russia is a technologically advanced country in the tradition of the former Soviet Union.[29]

The Russian response may indicate more widespread problems in Russian science and a challenge for the gradual reintegration of Russia into the international scientific community. Following the Russian invasion of Ukraine, virtually all the leading pharmaceutical companies reacted with a series of measures that amounted to de facto sanctions against the Russian Federation.[30] Not since World War II has a major power been isolated from the global research community in the medical sciences. In principle, research with human participants should be performed in places where societal oversight is possible through the instruments provided by democracy, including the free flow of information between scientific organizations, their members, and the media. However, some of the locations most in need of new interventions may be marked by local corruption. Unless the corruption is so pervasive as to make it impracticable to obtain valid results that could provide benefits for the population, it is thought to be inappropriate to discriminate among venues for clinical studies based on their national governance models alone. Indeed, even during the Cold War, clinical trials in the Eastern European communist states seem to have operated consistently with the conventions of the day. The German historian of medicine Volker Hess has documented hundreds of studies conducted in the former East Germany under the sponsorship of Western drug companies.[31] In the late 1950s, the United States approved the Sabin polio vaccine following testing in the Soviet Union involving millions of children.[32]

Consistent with international conventions within the rules-based order, for humanitarian reasons major drug companies have continued to supply "essential medicines" to Russia, a list compiled by the WHO as "those that satisfy the priority health needs of a population."[33] A UN Security Council resolution adopted in 2022 affirmed that sanctions should not interfere with the support of "basic human needs;"[34] this was intended partly to respond to concerns by bankers that sanctions could interfere with related financial activities. This is a notable humanitarian success of the rules-based order in responding to a critical and complex crisis. (International sanctions have however complicated delivery of many medications on the WHO list.)[35] However, companies including MSD (Merck), Novartis, Roche, Astra Zeneca, and Johnson & Johnson appear to have largely suspended recruitment for clinical trials. This situation has the look of an implicit sanction on the part of major drug companies, but it may just as well be a reluctance to enter into uncharted territory in the already complex domain of human experiments. There is also an element of self-isolation in Russian science policy. Not long after the invasion was frustrated by Ukrainian forces and sanctions began to be imposed, the Russian Ministry of Science and Higher Education informed scientific institutions that they need not expect their workers to index their publications in the Web of Science and Scopus.[36]

Long-term isolation from the international scientific community could give reasons to doubt Russia's compliance with international norms. Even by the nation's own standards, there are reasons to believe that Russia's record of compliance has been poor. Clinical pharmaceutical studies in Russia are supposed to be governed by the Central Ethics Committee of the Ministry of Health; regulated by the Federal Law No. FZ-61 (April 12, 2010), "On the Circulation of Medicinal Products;" and the Ministry of Health's Order 200n, "On Approval of Rules for Good Clinical Practice." According to these regulations, recruitment must be "aimed at . . . respecting the rights of participants," protecting their health, and verifying "the scientific validity of the study based on non-clinical tests and previous clinical investigations."[37] However, an investigation carried out in 2021 by members of the Russian Academy of Sciences Commission for Counteracting the Falsification of Scientific Research found that twenty-two out of twenty-six clinical trials had done one or more of the following: published articles that did not comply with the established mandatory rules for clinical trials authorization; conducted studies in which the investigators manifested methodologies that did not correlate with related publications; or reported that the "investigational product was efficacious, despite there being no statistically significant difference between the compared treatment arms." The Russian Academy of Sciences commission concluded that "publications of clinical study results in Russian journals frequently do not meet common standards and contain large evidence of ethical, scientific, and publishing breaches."[38]

These concerns were deepened during the research that the Russian Direct Investment Fund, the Kremlin's sovereign wealth fund, sponsored for the development of the COVID-19 Sputnik V vaccine. In 2021, the European Medicines Agency initiated a probe into whether those who were enrolled for the clinical trials involving Sputnik V, most of them military service members and civil servants, could have been coerced for participating.[39] Apart from possible violations in the ethics of subject recruitment and prior unexplained discrepancies in reported findings, there are also reasons to doubt the results of Russian COVID-19 vaccine studies. In September 2020, the *Lancet* published the findings of phases I and II of the Sputnik clinical trial. According to the paper, all the seventy-six participants in an open, nonrandomized trial "developed SARS-CoV-2 antibodies," with no serious adverse events detected.[40] Subsequently, in February 2022, the *Lancet* published another article, this time with the results of an interim phase III trial of Sputnik V, in which 22,000 adult participants received either two doses of the vaccine or placebo, demonstrating 91.6% effectiveness in preventing infection in the active arm and no serious adverse events.[41] However, the optimism reflected in the work published by the *Lancet* was refuted by a large group of researchers and experts in scientific integrity who pointed out serious methodological inconsistencies that would require analysis of the study's source data, which, so far, have not been presented for scrutiny.[42] As of this writing, these concerns have not been addressed.

While it is by no means the only offender, the history of human experiments in Russia during the Cold War has yet to be written. Recent archival work has found that there was extensive medical research in the Stalinist Soviet prison camps, or gulags.[43] Notably, the Nuremberg Code was not published in Russian until 1986, nor were Soviet scientists well informed about the World Medical Association's Declaration of Helsinki. According to one Russian scholar, "Probably, the most important reason to hide the Nuremberg trial and Code from Soviet physicians was to make physicians morally blind in relation to experiments they were appointed to do."[44]

All this leads to the question whether there has been an underlying cultural weakness in the ethical conditions for clinical trials in Russia, conditions exacerbated by the political pressures of the invasion and the pandemic. Part of the fault may lie in an underdeveloped research ethics culture in Russia. We have seen that the Nuremberg Code had little impact in the US until the domestic research ethics scandals of the 1960s and 1970s, although the trial was reasonably well publicized in the US. As I have described, historians have noted that the principal American experts at the trial were keen to protect "normal" medical science from being identified with the Nazi crimes. Much the same must be said of the international medical profession's foot-dragging in the years before the first version of the Declaration of Helsinki.

Although legitimate questions can be raised about the clinical trials sector, more generally there are examples of Russian scientists' concern about defending the integrity of science from politics. At a meeting of the Presidential Council for Human Rights held in the Kremlin in 2017, Vladimir Putin stated that "they [foreigners] purposefully and professionally" are collecting biological material about people of different ethnic groups and individuals residing in various geographical locations of the Russian Federation. Oleg Balanovsky, Head of the Laboratory of Genomic Geography of the Vavilov Institute of General Genetics, responded that the potential of using the "collected" biomaterials for harmful purposes is unfounded. Soon after, the Bulletin "In Defense of Science," a publication of the Commission for Combating Pseudoscience and Falsification of Scientific Research under the Presidium of the Russian Academy of Sciences, included a commentary on Putin's statement about developing biological weapons targeting specific ethnic groups. The Commission expressed its deep concerns about such suggestions and presented an expert opinion that creation of such "ethnic biological weapons," referring to a system of genetic approaches allegedly capable of causing the death of certain peoples without affecting others such targeting the Russian population, is impossible. The Commission continued with the more obvious argument to a lay audience: Biological population groups do not correspond to either states or peoples, as boundaries between populations have formed over millennia, and modern national borders have been defined by recent history. An individual's belonging to an ethnic group, at this point in the history of population migration, is highly determined by their culture and self-identity, not just the DNA code.[45]

In 2022, a Russian specialist in bioinformatics at the Skolkovo Institute of Science and Technology Mikhail Gelfand deviated from a regular Zoom lecture on comparative genomics to urge protests against the invasion of Ukraine. He also initiated a protest letter that quickly drew eight thousand signatories from the ranks of established Russian academicians. If Russia is turned into a pariah state, the letter declared, "this means that we scientists will no longer be able to do our job normally: after all, conducting scientific research is unthinkable without cooperation with colleagues from other countries. The war with Ukraine is a step to nowhere."[46]

Bioethics and the "Nylon Curtain"

As Karolina Zhukoff has argued, Western bioethicists and international institutions "can strengthen the global bioethics landscape and create pathways for Russian scientists to challenge unethical policies, ensuring that bioethical standards are upheld, even in authoritarian settings."[47] During the Cold War bioethics had at times a convening role with a specific human rights orientation during that decades-long superpower standoff. Joseph J. Fins describes one such occasion:

> In the late 1980s, the Hastings Center hosted scholars from Eastern Bloc countries and the former Soviet Union. They were there as bioethics scholars, but, in fact, they were there as dissidents and champions for democracy and civil liberties in their countries. The theory was if patients could get rights in the hospital with the recognition of their autonomy and self-determination, citizens could get rights in the streets and in civil society. In Eastern Europe and the former Soviet Union, bioethics was a crypto-democratic movement. It was much more than making choices about a ventilator at the end of life—it was about making choices about how one wanted to live one's life.[48]

As this anecdote suggests, during the Cold War, scientific, cultural, and intellectual exchanges were far more frequent than the "Iron Curtain" cliche. The East-West divide of the era was better described as a "Nylon Curtain," a term that historians have come to use to signify that, despite attempts at isolation from the West, the Soviet Union and the Eastern European countries continued to be part of a European culture.[49] Recent scholarship has shown that ideas, musical forms, and even some consumer goods found their ways east, along with the occasional (actual) nylons and blue jeans smuggled in by Western hippies for lucrative sales. Tourism went through phases and varied by country, as did academic travel, including visits to American bioethics centers.

This anecdote also illustrates the fact that bioethics is a field embedded in the world produced (or at least the narrative created) by the rules-based order, and particularly the discourse about human rights. It's not difficult to understand why a principle of respect for autonomy that emerged from a long series of scandals and controversies about human experiments dovetailed so well with the development of human rights law since 1945, nor how important the insistence on worrying about human rights has been in the post–World War II era. However, with some significant exceptions, bioethics discourse has largely nibbled at the edges of the domain of human rights. A diminished rules-based order may force a more intense engagement. In words of the Hungarian lawyer-ethicist Judit Sándor, "Human rights and bioethics are two distinct disciplines with different methods and scope. . . . Nevertheless," she continues, "these two fields have grown closer over the last couple decades for at least two new developments. This transformation has been driven partly by the increasing institutionalization of bioethics: with the establishment of bioethics committees and the legal formulation of procedures, codes, declarations, and even conventions. On the other hand, bioethics not only applies human rights terms and norms, but also provides useful tools for the analysis of human rights questions."[50]

Consider then, what it would mean for bioethics to be shorn of the background support of the universality of a rules-based order with its standards, norms, and conventions. A 2025 WHO initiative aims to develop a consensus-based "clinical ethics guidance" through an international working group of more than two dozen members drawn from Africa, Asia, Oceania and North and South America. The resulting guidance "will provide comprehensive direction on the governance of ethical issues in clinical practice, ensuring that ethical considerations are effectively integrated into healthcare systems globally."[51] This ambitious effort embodies the values of bioethics in the rules-based order. What if previously common standards endorsed by the post–World War II consensus of sovereign states no longer undergird even a plausible aspirational consensus? The situation would recall the traditional position of medical professionalism in which practitioners and investigators would then be on their own, largely out of reach of government regulators and the legal system. Preservation of human rights, and in particular the rights of patients, participants, and the public would depend on the invisible college assuming a visible form. In that case states need to take note of the norms asserted by medical science practitioners and scientists. As Jonathan Marks observes, "The legitimacy of the state depends, in part, on the assistance of professionals in the performance of the state's human rights obligations. Put simply, the legitimacy of the states depends on professionals."[52] The role of medicine in the Third Reich is an example of the extreme burden incurred by professional ethics in the absence of a human rights regime recognized by the state: The enthusiastic cooperation of the medical profession both legitimized and finally operationalized the Nazi program.

A New Ethics of Public Health

Assume, however, that we are not faced with such a grim prospect, that somehow despite the dark turn of recent US domestic politics reason will prevail, even if born out of national self-interest. Rehabilitating confidence in global public health institutions after the COVID-19 pandemic will nonetheless require coming to terms with the fact that, for practices like masking and lockdowns, respect for autonomy and its implementation were in tension with the mission of preserving population health. Some bioethicists have sought to reframe the basis of public health as social justice,[53] or to argue that health and human rights are "synergistic."[54] Here, too, entities operating within the rules-based order have laid down a marker for justifiable restrictions on human rights. In 1985, an international coalition of jurists proclaimed the Siracusa Principles, including the core proposition that "in applying a limitation [on rights], a state shall use no more restrictive means than are required for the achievement of the purpose of the limitation," and further that public health measures "must be specifically aimed at preventing disease or injury or providing care for the sick and injured."[55] The Siracusa Principles, developed in the early years of the HIV/AIDS pandemic, have ever since been a reference point for assessing public health restrictions on individual liberties.

Philosopher and bioethicist Nick Evans has warned against the tendency to "securitize" public health as a "war," with all its implications of self-defense and threat analysis, rather than as justice in health security.[56] During the COVID-19 pandemic, resistance to measures like masking and minimizing contact grew out of the apparent opposition between personal preferences and the common good. Granting the broad applicability of the Siracusa Principles, how can those acute disputes be resolved? One approach justifies infringements on moral considerations, like respect for autonomy, in cases where certain other conditions are obtained: such as the effectiveness of the policy, that any infringements of other moral considerations are proportionate to the public health benefits, that there is no less troubling policy available, that the policy minimizes infringement on other moral considerations, and that the policy can be publicly justifiable.[57] Another approach recognizes that the principle of respect for autonomy, along with the other standard moral principles of nonmaleficence, beneficence, and justice, has been developed for the research and clinical settings and is ill-suited to public health. To justify constraints on those moral considerations, this approach seeks to develop a framework of principles that grows from public health practice, wherein the concern is for the health of a population rather than for the individual patient. Ross Upshur has proposed four such principles, three of which are roughly similar to several in the previous analysis: minimizing harm, the least restrictive means, and transparency.[58]

Embodied in the principle of respect for autonomy, an individual rights orientation was richly justified by the scandals and tragedies so familiar to students of bioethics and has been embedded in the rules-based order. Yet, the COVID-19 pandemic put on vivid display the limitations of that tradition amid a distinct absence of solidarity, the "social feeling" that Baker describes. In a related point, Baker argues that the Belmont principle of respect for persons rather the Beauchamp and Childress principle of respect for autonomy would have made for a stronger bioethical analysis for the pandemic, such that "bioethicists would have had better conceptual tools for confronting the issues of vaccine hesitancy and mask resistance."[59] Indeed, as he notes, an autonomy principle simpliciter has been too easily hijacked in politicized arguments against modest public health considerations for the sake of "personal choice." The point here is not that improved bioethical analysis could have changed the outcome, but at least it wouldn't have contributed so much to autonomy run amuck.

A fourth public health ethics principle that Upshur identifies, "reciprocity," is largely absent from the literature on research and clinical ethics. (A notable exception is a 2013 UK Government guidance on pandemic flu.)[60] In part, this is because reciprocity depends on a set of background conditions that are not found in the sociopolitical setting of conventional American bioethics. These conditions generally fall under the heading of solidarity, a concept more familiar in Europe and Asia than in the US. Nor does the principle of justice fully capture the elements of "group obligation, mutuality and social feeling," as Baker has written. He continues: "Europeans believe that it is an important principle essential to counterbalancing the dispassionate claims of justice and the anomic forces of individualism and market capitalism (that they see embedded in Beauchamp and Childress's principle of autonomy)." Nor is convergence between a principle of respect for autonomy and the Asian concept of family autonomy plausible, Baker argues, as a standard interpretation of principlism justifies "clinical intrusion on family autonomy."[61] In defense of principlism, one might argue that reciprocity is implicit in any framework of moral principles, and in particular that the same practical consequences of mutual consideration can be teased out of the principles as conventionally presented. But in the postpandemic world such an exercise seems not to meet the moment.

The bioethicist Angus Dawson goes still further, I think rightly, as he urges "resetting the parameters" for a substantive public health ethics. After considering various versions of the approaches considered above, he observes that they tend to presuppose individual liberty, thus failing "to capture the *public* nature of public health" (emphasis in original).[62] His construction of a theory of the ethics of public health draws on a conception of the goods, harms, and interests human beings hold in common, those that are required for human flourishing that represent our convergent and common interests. "Individuals," Dawson writes, "are neither just individuals nor merely constituent parts of groups, societies, communities and populations. What we need is a public health ethics, and in turn a bioethics, that is both sensitive to and grounded in the reality of relevant social relationships."[63]

Green Horizons

In the late 1960s and early 1970s, right around the time that modern bioethics began to form its now recognizable set of themes and concerns, two other closely related topics attracted both expert and media attention. One was population growth, reviving Malthusian worries about an unsustainable rate of increased global resource consumption. "The Limits to Growth" was the title of a Club of Rome–sponsored study about the dynamics of human-earth systems, the results of which were presented in Moscow and Rio de Janeiro.[64] Vastly improved agricultural practices, known collectively as the Green Revolution, have staved off the most severe consequences predicted by the Club of Rome report, such that food insecurity is now mainly a result of inadequate distribution systems. However, other factors now threaten access to both food and water for much of the world. A second topic that was prevalent in those years was environmental protection, an international movement led by UNESCO that culminated in the first Earth Day in 1970. Population and environment were linked by fears of ecological ruin due to irresponsible human activity. The Anthropocene Age is a popular modern framing of the irreversible consequences for the planet of certain human actions, such as the geological impact of mining for fossil fuel. Today the movement to mitigate climate change has brought these threads together.

When modern medical ethics needed a term to capture its departure from the tradition of physician-oriented codes, "bioethics" was in the air. University of Wisconsin chemist Van Rensselaer Potter was then using the word to refer to a concept that was more global than medical ethics, an integration of biology and human values.[65] Nonetheless, it is striking that all of these ideas were floating around in that period, such that the new field landed on a name laden with so many closely related ideas unified by a common theme of human beings as embedded in nature, not only in but also of a delicate balance of forces. A twenty-first-century bioethics that reflected its implicit implications would do well to broaden its horizons such that *bio-* is given its full force and a principle of justice given a more fundamental role in unfolding the implications of bioethical commitments.

The familiar rules-based order within which bioethics has flourished cannot claim to have managed the climate threat all that well. The hegemony of powerful fossil fuel consumers like the United States justifies the resentment of less-developed states that are being asked to forego the advantages that others have exploited. But it is hard to see how a diminished world order, or one that is reorganized into some new set of alliances and divisions, could do better. A new order in which the effects of climate change challenge the future of human civilization will inevitably engage bioethical issues and arguments. Nor is climate change necessarily the most immediate threat to human civilization when one considers that large regional wars in both Europe and the Middle East have never before broken out simultaneously in a multipolar world stocked with nuclear weapons.

Coda

I have argued that the consent standard that operationalizes a principle of respect for autonomy has exercised a firm hold on bioethics–American style since the origins of the rules-based international order. Consent presupposes a person, a sovereign individual capable of moral agency. The most insistent case was that of the experimental subject, then the compelling moral argument for a consent standard gradually made its way into the clinical setting in theory in much of the world, and over the years with often needful, sometimes awkward, applications in public health and in the ethics of basic science.

The international rules-based order defined by open markets, liberal democracy, and multilateral organizations was the aspirational, if not always actual or evenhanded, basis for the post–World War II vision, a burden that now falls on those who would propose a plausible alternative. The struggle for a new global order will be the principal geopolitical agenda for at least the second quarter of the twenty-first century, bringing with it new challenges for bioethics in theory and in practice. In conjunction with machine learning and cloud computing becoming accessible to all, the extraction of information from massive data sets appears to be an inevitable and inescapable part of the new global bioethics landscape, a bioethics of data science. Consider for example the "data subject," understood very broadly in the EU's 2018 General Data Protection Regulation (GDPR) as an identified or identifiable natural person.[66] Under this rubric, intended "to ensure the free movement of data throughout the European Union" and the conditions of its transfer outside the EU, any data that does or could identify a person is personal data.[67] The European Union is often seen as the world's "regulation cop," a familiar epithet amid vigorous objections and warnings about the threat to the free flow of information that the GDPR represents. Nonetheless, the EU has for years been negotiating terms for data transfer with other leading powers like the US and Japan, effectively assuming a leadership role in diffusing its influence through the multiplicity of global institutions. The very language of data privacy used in the EU has taken hold in corporate America.[68]

In our twenty-first-century cyberworld, data collection standards and practices cut across all four of my somewhat simplistic classifications of the subfields of bioethics—research ethics, clinical ethics, the ethics of public health, and the ethics of basic science—therefore they present a unifying challenge for the field. One example is the concept of the "learning health system" in which data from countless patients is part of a feedback loop with aggregated information from countless other patients, ultimately yielding improved treatment.[69] A learning health system not only begins to collapse the classical distinction between research and therapy, it also potentially embraces public health and basic science data, all aimed finally at patient care.[70] Everyone fortunate enough to be included in data collection for the common good will be a data subject, although many will be excluded or exploited.

Besides familiar obstacles to benefit sharing, if the transition to a novel data protection regime continues to unfold there will be plenty of bioethical problems to chew on. For example, in an attempt to deal with the possibility that excessive data protection can discourage scientific exchange (recall the invisible college), the GDPR allows exceptions for research conducted under "recognized ethical standards," but those standards might allow for broad consent in conflict with the requirements imposed by research ethics committees or national regulations. An example is the use of biobank materials for genetics research, in which GDPR conditions–like notification of participants, their right to move their data between data controllers ("portability"), and the right to object to continued research–would be hard to harmonize and challenging to enforce, undermining the goal of the regulations.[71]

Then there are the usual problems with any AI-based system. The rapid development and application of facial recognition technologies (FRT) for personal identification exemplifies both technical limitations like biased data sets as well as the challenge of harmonization between legal and political systems. In theory FRT could vastly ease essential activities like commerce and travel, but it has already been implicated in the growth of a surveillance state beyond legitimate policing, despite the current systems' technical flaws, much less human rights concerns.[72] Consider in this context the language of the GDPR's article 23 which refers to "fundamental rights and freedoms" in the context of safeguards for a "democratic society." These are the very same conditions presupposed by the post–World War II rules-based order, that any constraints on liberty are only justifiable within the requirements and for the preservation of a liberal democracy.

Reminiscent of the 1930s, but not, one hopes, with such a catastrophic climax, the tectonic plates of the international order are grinding and shifting, presenting novel challenges for global institutions, bioethics among them. I began this book asserting that bioethics is a creature of the rules-based international order, and that has been the burden of my argument. In the course of writing I have come to believe that one could go further: Bioethics is also paradigmatic of the rules-based order as we have known it, for all its limitations and failures. In the face of global existential risks and lacking a reasonable alternative, an ordered international system of liberal nation-states under recognized norms and conventions remains absolutely essential.

Glossary

AEC: Atomic Energy Commission

CCNE: Comite Consultatif National D'Ethique

CIA: Central Intelligence Agency

CIOMS: Council for International Organizations of Medical Science

Codex: Codex Alimentarius

DoD: Department of Defense

DoH: Declaration of Helsinki

DSMB: Data Safety Monitoring Board

FAO: Food and Agriculture Organization of the United Nations

FDA: Food and Drug Administration

GATT: General Agreement on Tariff and Trade

GCP: Good Clinical Practices

GWOT: Global War on Terror

IAB: International Association of Bioethics

IAP: InterAcademy Partnership

IBC: International Bioethics Committee

ICH: International Conference on Harmonization

ICMMP: International Committee on Military Medical Pharmacy

ICRC: International Committee of the Red Cross

IMT: International Military Tribunal

IMTFFE: International Military Tribunal for the Far East

IRB/REB: Institutional Review Board/Research Ethics Board

LMIC: Low- and Middle-Income Countries

LOAC/IHL: Law of Armed Conflict/International Humanitarian Law

NEC: National Ethics Committee

NIH: National Institutes of Health

NMT: Nuremberg Military Tribunals

OECD: Organization for International Cooperation and Development

OMS: Office of Medical Services, Central Intelligence Agency

PHS: Public Health Service

UDBHR: Universal Declaration on Bioethics and Human Rights

UNESCO: United Nations Educational, Scientific and Cultural Organization

UNWCC: United Nations War Crimes Commission

WHO: World Health Organization

WMA: World Medical Association

Notes

Preface

1. I regard bioethics as an interdisciplinary field rather than as a discipline, hence my terminology in this book.

2. Doug Stokes, "Trump, American Hegemony and the Future of the Liberal International Order," *International Affairs* 94, no. 1 (January 2018): 133–50, https://www.chathamhouse.org/sites/default/files/images/ia/INTA94_1_8_238_Stokes.pdf.

3. G. John Ikenberry, "Why the Liberal World Order Will Survive," *Ethics & International Affairs* 32, no. 1 (Spring 2018), https://www.ethicsandinternationalaffairs.org/journal/why-the-liberal-world-order-will-survive.

4. Anya Wahal, "On International Treaties, the United States Refuses to Play Ball (Blog)," *Council on Foreign Relations* (blog), January 7, 2022, https://www.cfr.org/blog/international-treaties-united-states-refuses-play-ball.

5. Associated Press, "The Geneva Conventions—The World's Oldest Rules of War—Are 75 Years Old and Ignored Nearly Everywhere," August 12, 2024, https://apnews.com/article/international-committee-red-cross-geneva-conventions-ukraine-gaza-sudan-syria-438e5a2ecd8c34b5d1c810dbd7329aed.

6. Hugh Wilford, *The CIA: An Imperial History* (Basic Books, 2024).

7. A classic history is David J. Rothman, *Strangers at the Bedside: A History of How Law and Bioethics Transformed Medical Decision Making* (Basic Books, 1992). For a more recent account that also considers certain substantive bioethics topics, see Amy Gutman and Jonathan D. Moreno, *Everybody Wants to Go to Heaven but Nobody*

Wants to Die: Bioethics and the Transformation of Health Care in America (Liveright/ Norton, 2019).

8. By far the most influential treatment is Tom L. Beauchamp and James F. Childress, *Principles of Biomedical Ethics*, 8th ed. (Oxford University Press, 2019).

Chapter 1

1. Robert Baker, "Bioethics and Human Rights: A Historical Perspective," *Cambridge Quarterly of Healthcare Ethics* 10, no. 3 (June 29, 2001): 241–52, https://doi.org/10.1017/S0963180101003048.

2. G. John Ikenberry, *A World Safe for Democracy: Liberal Internationalism and the Crises of Global Order*, 1st ed. (Yale University Press, 2020), 2.

3. Ikenberry, *A World Safe for Democracy*, 2.

4. Derek Croxton, "The Peace of Westphalia of 1648 and the Origins of Sovereignty," *The International History Review* 21, no. 3 (1999): 569–82, https://doi.org/10.1080/07075332.1999.9640869.

5. David Chandler, "International Justice," *New Left Review* no. 6 (December 2000), https://newleftreview.org/issues/ii6/articles/david-chandler-international-justice.

6. Chandler, "International Justice."

7. Ikenberry, *A World Safe for Democracy*.

8. Pauline Kleingeld, "Approaching Perpetual Peace: Kant's Defence of a League of States and His Ideal of a World Federation," *European Journal of Philosophy* 12, no. 3 (December 2004): 304–25, https://doi.org/10.1111/j.0966-8373.2004.00212.x.

9. John Rawls, *The Law of Peoples* (Harvard University Press, 1999), 27.

10. "What Is R2P?," Global Centre for the Responsibility to Protect, accessed December 4, 2024, https://www.globalr2p.org/what-is-r2p/.

11. Robert H. Ferrell, *Peace in Their Time: The Origins of the Kellogg-Briand Pact* (New Haven: Yale University Press, 1952).

12. Julie M. Bunck and Michael R. Fowler, "The Kellogg-Briand Pact: A Reappraisal," *Tulane Journal of International & Comparative Law* 27, no. 2 (2019): 229–76.

13. John Lewis Gaddis, "The Long Peace: Elements of Stability in the Postwar International System," *International Security*, Spring 1986 (Vol. 10, No. 4),106.

14. Robert L. Beisner, *Dead Acheson: A Life in the Cold War* (Oxford University Press, 1966), 21.

15. Zachary D. Carter, *The Price of Peace: Money, Democracy, and the Life of John Maynard Keynes* (Random House, 2020).

16. Ikenberry, *A World Safe for Democracy*, 173.

17. Ikenberry, 179–80.

18. Gaddis, "The Long Peace," 99–100.

19. John J. Mearsheimer, *The Great Delusion: Liberal Dreams and International Realities*, The Henry L. Stimson Lecture Series (Yale University Press, 2018), 1.

20. Mearsheimer, *The Great Delusion*, 3.

21. Randall Lesaffer and Jane E. Nijman, eds., *The Cambridge Companion to Hugo Grotius* (Cambridge University Press, 2021), https://doi.org/10.1017/9781108182751.

22. Louis Henkin, "The Universality of the Concept of Human Rights," *The Annals of the American Academy of Political and Social Science* 506, no. 1 (November 1989): 10, https://doi.org/10.1177/0002716289506001002.

23. Judit Sándor, "Bioethics as a Tool for Extending the Human Rights Discourse," in *Human Rights in the 21st Century Amazon*, eds. Tibor Várady and Miodrag Jovanovic (Eleven International Publishing, 2020), 348.

24. Bryan H. Druzin, "Why Does Soft Law Have Any Power Anyway?," *Asian Journal of International Law* 7, no. 2 (2017): 361, https://doi.org/10.1017/S2044251316000229.

25. InterAcademy Partnership, *Doing Global Science: A Guide to Responsible Conduct in the Global Research Enterprise* (Princeton University Press, February 2016), https://www.interacademies.org/publication/doing-global-science-guide-responsible-conduct-global-research-enterprise.

26. InterAcademy Partnership, *Doing Global Science*, 363.

27. Henk ten Have, "Globalizing Bioethics Through, Beyond, and Despite Governments," in *Global Bioethics: The Impact of the UNESCO International Bioethics Committee*, eds. Alireza Bagheri, Jonathan D. Moreno, and Stefenao Semplici (Springer, 2016).

28. Nobu Hayashi, "Basic Principles," in *Routledge Handbook of the Law of Armed Conflict*, eds. Rain Livoja and Tim McCormack (2016), 721.

29. ICRC Database, Customary IHL, Practice relating to Rule 15, "Principle of Precautions in Attack," accessed March 05, 2023, https://ihl-databases.icrc.org/en/customary-ihl/v2/rule15.

30. Jann Kleffner, "Sources of the Law of Armed Conflict," in *Routledge Handbook of the Law of Armed Conflict*, eds. Rain Livoja and Tim McCormack (Routledge, 2016), 721.

Chapter 2

1. Charter of the International Military Tribunal, *Nuremberg Trial Proceedings Volume 1*, (Nuremberg, November 19, 1945), https://avalon.law.yale.edu/imt/imtconst.asp.

2. Telford Taylor, *The Anatomy of the Nuremberg Trials: A Personal Memoir* (Knopf, 1992), 5.

3. Kellogg-Briand Pact of 1928, https://avalon.law.yale.edu/20th_century/kbpact.asp.

4. Frits Kalshoven, "The History of International Humanitarian Law Treaty-Making," in *Routledge Handbook of the Law of Armed Conflict*, eds. Rain Livoja and Tim McCormack (Routledge, 2016), 721.

5. Royal Institute of International Affairs, *United Nations Documents 1941–1945* (Oxford University Press, 1946).

6. Francine Hirsch, *Soviet Judgment at Nuremberg: A New History of the International Military Tribunal after World War II* (Oxford University Press, 2020), 20–21.

7. Hirsch, *Soviet Judgment at Nuremberg*, 21.

8. Hirsch, *Soviet Judgment at Nuremberg*, 39.

9. Taylor, *The Anatomy of the Nuremberg Trials*, 39.

10. Hirsch, *Soviet Judgment at Nuremberg*, 316.

11. Howard Brody et al., "United States Responses to Japanese Wartime Inhuman Experimentation after World War II: National Security and Wartime Exigency," *Cambridge Quarterly of Healthcare Ethics*, no. 23 (April 2014): 2, https://doi.org/10.1017%2FS0963180113000753.

12. Gary Bass, *Judgment at Tokyo: World War II on Trial and the Making of Modern Asia* (Penguin Random House, 2023), 680.

13. Taylor, *The Anatomy of the Nuremberg Trials*, 42.

14. Ulf Schmidt, *Justice at Nuremberg: Leo Alexander and the Nazi Doctors' Trial* (Palgrave Macmillan, 2004), 120–22.

15. Hirsch, *Soviet Judgment at Nuremberg*, 416.

16. Schmidt, *Justice at Nuremberg*, 123.

17. Hirsch, *Soviet Judgment at Nuremberg*, 201.

18. Schmidt, *Justice at Nuremberg*, 203.

19. Kevin Jon Heller, *The Nuremberg Military Tribunals and the Origins of International Criminal* (Cambridge: Oxford University Press, 2011), 53.

20. For Alexander's biographical details discussed in this section see Schmidt, *Justice at Nuremberg*, chapter 3.

21. Schmidt, Justice at Nuremberg, 134.

22. Schmidt, *Justice at Nuremberg*, 137.

23. Jonathan D. Moreno, *Undue Risk: Secret State Experiments on Humans*, 1st ed. (Routledge, 2001), 74–79.

24. Schmidt, *Justice at Nuremberg*, 203.

25. Schmidt, *Justice at Nuremberg*, 203.

26. Schmidt, *Justice at Nuremberg*, 67.

27. Schmidt, *Justice at Nuremberg*, 163.

28. Schmidt, *Justice at Nuremberg*, 166.

29. Schmidt, *Justice at Nuremberg*, 144.

30. Heller, *The Nuremberg Military Tribunals and the Origins of International Criminal*, 87.

31. Schmidt, *Justice at Nuremberg*, 175.

32. Robert N. Proctor, *Racial Hygiene: Medicine Under the Nazis* (Harvard University Press, 1988).

33. Schmidt, *Justice at Nuremberg*, 207.

34. Katja Schmidt-Mai and Jonathan D. Moreno, "A Wartime Medical Experiment as Propaganda: The Malaria Case," in *Propaganda and Conflict: War, Media and Shaping the Twentieth Century*, eds. Mark Connelly et al. (Bloomsbury Academic, 2019).

35. International Conference on Military Trials: London 1945, "Revision of American Draft of Proposed Agreement, June 14, 1945," https://avalon.law.yale.edu/imt/jack09.asp.

36. Charles Garraway, "Superior Orders and the International Criminal Court: Justice Delivered or Justice Denied," *International Committee of the Red Cross* 81, no. 836 (1999): 785–94, https://international-review.icrc.org/sites/default/files/S1560775500103712a.pdf.

37. Jeanne Guillemin, *Hidden Atrocities: Japanese Germ Warfare and American Obstruction of Justice at the Tokyo Trial*, 1st ed., A Nancy Bernkopf Tucker and Warren I. Cohen Book on American–East Asian Relations (Columbia University Press, 2017), 281.

38. Daniel Barenblatt, *A Plague upon Humanity: The Secret Genocide of Axis Japan's Germ Warfare Operation* (HarperCollins, 2004), 222.

39. Moreno, *Undue Risk*, 16–20.

40. Friedrich H. Moll, Matthias Krischel, and Heiner Fangerau, "Albert Neisser and the First Prussian Directive on Informed Consent," in *Skeletons in the Closet: Indignities and Injustices in Medicine*, ed. Mike Moran (American Urological Association, 2012).

41. George J. Annas and Michael Grodin, eds., *The Nazi Doctors and the Nuremberg Code—Human Rights in Human Experimentation* (Oxford University Press, 1992).

42. Hans-Martin Sass, "Ambiguities in Judging Cruel Human Experimentation: Arbitrary American Responses to German and Japanese Experiments," *Eubios Journal of Asian and International Bioethics* 13 (2003): 102–4, https://www.eubios.info/EJ133/ej133g.htm.

43. Robert Dingwall and Vienna Rozelle, "The Ethical Governance of German Physicians, 1890–1939: Are There Lessons from History?," *Journal of Policy History* 23, no. 1 (January 2011): 29052, https://doi.org/10.1017/S0898030610000308.

44. Schmidt, *Justice at Nuremberg*; Robert Baker, *Making Modern Medical Ethics: How African Americans, Anti-Nazis, Bureaucrats, Feminists, Veterans, and Whistleblowing Moralists Created Bioethics* (MIT Press, 2024).

45. Jay Katz, "The Consent Principle of the Nuremberg Code: Its Significance Then and Now," in *The Nazi Doctors and the Nuremberg Code—Human Rights in Human Experimentation*, eds. George J. Annas and Michael Grodin (Oxford University Press, 1992).

46. Jenny Hazelgrove, "The Old Faith and the New Science: The Nuremberg Code and Human Experimentation Ethics in Britain, 1946–73," *Social History of Medicine* 15, no. 1 (April 2002): 133, https://doi.org/10.1093/shm/15.1.109.

Chapter 3

1. Jay Katz, *The Silent World of Doctor and Patient*, 1st ed. (Johns Hopkins University Press, 2002).

2. Claude Bernard, *An Introduction to the Study of Experimental Medicine* (Dover Publications, 1957).

3. Slater v Baker and Stapleton [1767] 95 Eng. Rep 860; 2 Wils. K.B. 359.

4. Robert Dingwall and Vienna Rozelle, "The Ethical Governance of German Physicians, 1890–1939: Are There Lessons from History?," *Journal of Policy History* 23, no. 1 (January 2011): 29–52, https://doi.org/10.1017/S0898030610000308.

5. Jonathan D. Moreno, Ulf Schmidt, and Steven Joffe, "The Nuremberg Code 70 Years Later," *JAMA* 318, no. 9 (2017): 795–96, https://doi.org/10.1001/jama.2017.10265.

6. J. A. Pridham, "Founding of the World Medical Association," *World Medical Association Bulletin* 3, no. 3 (July 1951): 207–8.

7. Robert Baker, *Making Modern Medical Ethics: How African Americans, Anti-Nazis, Bureaucrats, Feminists, Veterans, and Whistleblowing Moralists Created Bioethics* (MIT Press, 2024), https://mitpress.mit.edu/9780262547376/making-modern-medical-ethics/.

8. Jonathan D. Moreno, Judit Sándor, and Ulf Schmidt, "Autonomy and Social Responsibility: The Post-Pandemic Challenge," *Perspectives in Biology and Medicine* 65, no. 3 (Summer 2022): 426–41, https://doi.org/10.1353/pbm.2022.0035.

9. Moreno, Sándor, and Schmidt, "Autonomy and Social Responsibility."

10. 1959. évi 8. törvényerejű rendelet az orvosi rendtartásról, Tvr. 6 § (1).

11. French Ministry of Public Health and Population, "Decree No. 55–1591, Code of Medical Deontology," November 2, 1955, Articles 34, 63, and 64.

12. French Ministry of Public Health and Population, "Decree No. 55–1591, Code of Medical Deontology," November 28, 1955, Article 38.

13. Moreno, Sándor, and Schmidt, "Autonomy and Social Responsibility."

14. United Nations General Assembly, "Universal Declaration of Human Rights," A/RES/217(III) December 10, 1948. Preamble, Resolution 217A (III), https://www.un.org/en/about-us/universal-declaration-of-human-rights.

15. Ulf Schmidt, "Medicine's Vestal Virgins: The World Medical Association and the Making of International Medical Law," unpublished 2023, typeset.

16. Schmidt, "Medicine's Vestal Virgins," 1.

17. Schmidt, Ulf. "From Nuremberg to Helsinki: Historicizing the Codification of Post-War Research Ethics." In *Ethical Innovation for Global Health*, edited by Chieko Kurihara, Dirceu Greco, and Ames Dhai, 149–74. Springer Nature Singapore, 2023. https://doi.org/10.1007/978-981-99-6163-4_9.

18. Susan Lederer, "Research Without Borders," in *Twentieth Century Ethics of Human Subjects Research: Historical Perspectives on Values, Practices, and Regulations*, eds. Volker Roelcke and Giovanni Maio (Franz Steiner Verlag, 2004).

19. Schmidt, "From Nuremberg to Helsinki," 157–58.

20. Schmidt, 162.

21. Sev S. Fluss, "How the Declaration of Helsinki Developed," *Good Clinical Practice Journal* 6, no. 6 (January 1999): 18–22.

22. Schmidt, "From Nuremberg to Helsinki," 163.

23. Schmidt, 166.

24. Susan Lederer, "Research Without Borders," in *Twentieth Century Ethics of Human Subjects Research: Historical Perspectives on Values, Practices, and Regulations*, eds. Volker Roelcke and Giovanni Maio (Franz Steiner Verlag, 2004), 159.

25. I served on an Institute of Medicine (now National Academy of Medicine) committee that recommended further reforms in prison research: Lawrence O. Gostin,

Cori Vanchieri, and Andrew Pope, eds., *Ethical Considerations for Research Involving Prisoners* (The National Academies Press, 2007).

26. Pertinax, "Without Prejudice," *British Medical Journal* 1, no. 5344 (June 8, 1963).

27. Susan S. Ellenberg and Robert Temple, "Placebo-Controlled Trials and Active-Control Trials in the Evaluation of New Treatments," *Annals of Internal Medicine* 133, no. 6 (September 19, 2000): 464–70, https://doi.org/10.7326/0003-4819-133-6-200009190-00015.

28. R. K. Lie et al., "The Standard of Care Debate: The Declaration of Helsinki Versus the International Consensus Opinion," *Journal of Medical Ethics* 30, no. 2 (April 2004): 190–93, https://doi.org/10.1136/jme.2003.006031.

29. P. G. de Roy, "Helsinki and the Declaration of Helsinki," *World Medical Journal* 50, no. 1 (March 2004): 9–11.

30. Canadian Medical Association Journal Editors, "Dismantling the Helsinki Declaration," *Canadian Medical Association Journal* 169, no. 10 (November 11, 2003): 997, https://www.cmaj.ca/content/169/10/997.

31. Cited in Howard Wolinksy, "The Battle of Helsinki," EMBO Rep. 2006 Jul; 7 (7): 670–672.

32. Howard Wolinsky, "The Battle of Helsinki: Two Troublesome Paragraphs in the Declaration of Helsinki Are Causing a Furor Over Medical Research Ethics," *EMBO Reports* 7, no. 7 (July 2006): 670–72, https://doi.org/10.1038%2Fsj.embor.7400743.

33. Wolinsky, "The Battle of Helsinki".

34. Amy L. McGuire et al., "Ethical Challenges Arising in the COVID-19 Pandemic: An Overview from the Association of Bioethics Program Directors (ABPD) Task Force," *American Journal of Bioethics* 20, no. 7 (June 8, 2020): 15–27, https://doi.org/10.1080/15265161.2020.1764138.

35. Jonathan H. Marks, "COVID-19, Pandemic Triage, and the Polymorphism of Justice," *American Journal of Bioethics* 20, no. 7 (July 27, 2020): 103–6, https://doi.org/10.1080/15265161.2020.1779410.

36. Annette Rid and Harald Schmidt, "The 2008 Declaration of Helsinki—First Among Equals in Research Ethics?," *American Society of Law, Medicine, & Ethics* 1, no. 38 (March 1, 2010): 143–48, DOI: 10.1111/j.1748-720X.2010.00474.x

37. Food and Drug Administration, "Human Subject Protection; Clinical Studies Not Conducted Under and Investigational New Drug Application," April 8, 2008, https://www.federalregister.gov/documents/2008/04/28/E8-9200/human-subject-protection-foreign-clinical-studies-not-conducted-under-an-investigational-new-drug

38. Stuart Rennie, "The FDA and Helsinki," *The Hastings Center Report* 39, no. 3 (June 2009): 49, https://doi.org/10.1353/hcr.0.0135.

39. Peter Lurie and Dirceu B. Greco, "U.S. Exceptionalism Comes to Research Ethics," *Lancet* 365, no. 9465 (March 26, 2005): 1117–19, https://doi.org/10.1016/s0140-6736(05)71856-0.

40. Lie et al., "The Standard of Care Debate."

41. Chieko Kurihara, "Interview with Dr. Robert Temple on Drug Evaluation Policy of FDA: Ethics, Science of Placebo-Control and Comparative Effectiveness Studies," *Rinsho Hyoka (Clinical Evaluation)* 42 (2014): 539–51, http://cont.o.oo7.jp/42_2/p539-51eng.pdf.

42. L. J. Burgess and D. Pretorius, "FDA Abandons the Declaration of Helsinki: The Effect on the Ethics of Clinical Trial Conduct in South Africa and Other Developing Countries," *The South African Journal of Bioethics & Law* 5, no. 2 (2012): 1–6, https://www.ajol.info/index.php/sajbl/article/view/83499.

43. Michael D. E. Goodyear et al., "Does the FDA Have the Authority to Trump the Declaration of Helsinki?," *BMJ* 338 (April 21, 2009): b1559, https://doi.org/10.1136/bmj.b1559.

44. Volnei Garrafa and Dora Porto, "Bioethics, Power and Injustice: For an Ethics of Intervention," *Journal International de Bioethique* 14, no. 1–2 (June 2003): 25–40, 215–16, https://www.cairn.info/revue-journal-international-de-bioethique-2003-1-page-23.htm.

45. CIOMS, "Bioethics," https://cioms.ch/bioethics/.

46. Bryant, John H. "Poverty, Vulnerability, the Value of Human Life, and the Emergence of Bioethics." In *Poverty, Vulnerability, the Value of Human Life, and the Emergence of Bioethics: Highlights and Papers of the XXVIIIth CIOMS Conference,* Ixtapa, Guerrero State, Mexico, 17-20 April 1994, edited by John H. Bryant, Zbigniew Bańkowski, and Council for International Organizations of Medical Sciences. CIOMS, 1995.

47. World Health Organization, "Regulation and Prequalification," https://www.who.int/teams/regulation-prequalification/regulation-and-safety/pharmacovigilance

Chapter 4

1. Kayte Spector-Bagdady and Paul A. Lombardo, "U.S. Public Health Service STD Experiments in Guatemala (1946–1948) and Their Aftermath," *Ethics & Human Research* 41, no. 2 (April 2019): 29–34, https://doi.org/10.1002/eahr.500010.

2. Presidential Commission for the Study of Bioethical Issues, "'Ethically Impossible': STD Research in Guatemala from 1946 to 1948" (Government Printing Office, September 2011), https://bioethicsarchive.georgetown.edu/pcsbi/node/654.html.

3. John Marks, *The Search for the Manchurian Candidate: The CIA and Mind Control* (W.W. Norton, 1979).

4. Karen M. Masterson, *The Malaria Project: The U.S. Government's Secret Mission to Find a Miracle Cure* (Random House, 2014).

5. The Advisory Committee on Human Radiation Experiments, *Final Report of the Advisory Committee on Human Radiation Experiments* (Oxford University Press, 1996), 152.

6. The Advisory Committee On Human Radiation Experiments, *Final Report of the Advisory Committee*, 55.

7. Jonathan D. Moreno, *Undue Risk: Secret State Experiments on Humans* (W. H. Freeman, 1999).

8. Moreno, *Undue Risk*, 56.

9. Moreno, 91.

10. Scott Podolsky, *The Antibiotic Era: Reform, Resistance, and the Pursuit of a Rational Therapeutics* (Johns Hopkins University Press, 2015), 92.

11. Laura Stark, *Behind Closed Doors: IRBs and the Making of Ethical Research* (University of Chicago Press, 2011), 162.

12. Office of the Special Assistant to the Under Secretary of Defense (Personnel and Readiness) for Gulf War Illnesses, Medical Readiness and Military Deployments, "Project Shipboard Hazard and Defense (SHAD): Autumn Gold," https://www.health.mil/Reference-Center/Fact-Sheets/2003/12/02/Autum-Gold#:~:text=The%20Autumn%20Gold%20test%20used,the%20details%20of%20each%20test.

13. Jonathan D. Moreno, "Bioethics and the National Security State," *Journal of Law, Medicine, & Ethics* 32, no. 2 (Summer 2004): 198–208, https://doi.org/10.1111/j.1748-720x.2004.tb00466.x.

14. "Most Guantanamo Detainees Are Innocent: Ex Bush Official," Associated Press, March 19, 2009.

15. Fionnuala Ní Aoláin, *Technical Visit to the United States and Guantánamo Detention Facility by the Special Rapporteur on the Promotion and Protection of Human Rights and Fundamental Freedoms While Countering Terrorism* (United Nations, June 14, 2023), https://www.ohchr.org/sites/default/files/documents/issues/terrorism/sr/2023-06-26-SR-terrorism-technical-visit-US-guantanamo-detention-facility.pdf.

16. "Memoramdum for William J. Haye II, General Counsel of the Department of Defense, Re: Military Interrogation of Alien Unlawful Combatants Held Outside the United States," Office of Legal Counsel, U.S. Department of Justice, March 14, 2003, https://www.justice.gov/sites/default/files/olc/legacy/2009/08/24/memo-combatantsoutsideunitedstates.pdf.

17. "Committee Study of the Central Intelligence Agency's Detention and Interrogation Program," Select Committee on Intelligence, December 9, 2014, https://www.intelligence.senate.gov/sites/default/files/publications/CRPT-113srpt288.pdf.

18. "Summary and Reflections of Chief of Medical Services on OMS Participation in the RDI Program," Office of Medical Services, Central Intelligence Agency, August 14, 2018, https://www.aclu.org/publications/summary-and-reflections-chief-medical-services-oms-participation-rdi-program.

19. Scott A. Allen and Nathaniel Raymond, *Experiments in Torture* (Physicians for Human Rights, June 2010), https://phr.org/wp-content/uploads/2010/06/Experiments_in_Torture.pdf.

20. Jonathan Marks, personal correspondence to Jonathan D. Moreno, February 8, 2024.

Chapter 5

1. Donald J. Krogstad et al., "Informed Consent in International Research: The Rationale for Different Approaches," *American Journal of Tropical Medicine and Hygiene* 83, no. 4 (October 5, 2010): 743–47, https://doi.org/10.4269%2Fajtmh.2010.10-0014.

2. Fritz Jahr, "Bio-Ethik: Eine Umschau Über Die Ethischen Beziehungen Des Menschen Zu Tier Und Pflanze [Bio-Ethics: A Review of the Ethical Relationships of Humans to Animals and Plants]," *Kosmos* 24 (1927): 2–4.

3. Cited in Hans-Martin Sass, "Fritz Jahr's 1927 Concept of Bioethics," *Kennedy Institute of Ethics Journal* 17, no. 4 (December 2007): 279–95, https://doi.org/10.1353/ken.2008.0006.

4. Van Rensselaer Potter, "Bioethics, The Science of Survival," *Perspectives in Biology and Medicine* 14, no. 1 (Autumn 1970): 127–53, https://doi.org/10.1353/pbm.1970.0015.

5. UNESCO, "Universal Declaration on Bioethics and Human Rights," 19 October 2005, https://www.unesco.org/en/legal-affairs/universal-declaration-bioethics-and-human-rights.

6. D. J. McCormack, A. Gulati, and J. Mangwani, "Informed Consent: A Global Perspective," *The Bone and Joint Journal* 100-B, no. 6 (June 2018): 687–92, https://doi.org/10.1302/0301-620x.100b6.bjj-2017-1542.r1.

7. National Commission for the Protection of Human Subjects of Biomedical and Behavioral Research, *The Belmont Report: Ethical Principles and Guidelines for the Protection of Human Subjects of Research* (Department of Health, Education, and Welfare, April 18, 1978).

8. Donald Oken, "What to Tell Cancer Patients a Study of Medical Attitudes," *JAMA* 175, no. 13 (April 1, 1961): 1120–28, https://doi.org/10.1001/jama.1961.03040130004002.

9. Dennis H. Novack, Robin Plumer, and Raymond L. Smith, "Changes in Physicians' Attitudes Toward Telling the Cancer Patient," *JAMA* 241, no. 9 (March 2, 1979): 897–900, https://doi.org/10.1001/jama.1979.03290350017012.

10. Jay Katz, *The Silent World of Doctor and Patient*, 1st ed. (Johns Hopkins University Press, 2002).

11. Jonathan D. Moreno, personal correspondence to Harald Schmidt, "Meeting Note/HS—Proposal for a *Declaration on Universal Norms on Bioethics* International Bioethics Committee (IBC) of UNESCO, Paris, 27–29 April 2004"

12. "Universal Declaration of Human Rights," United Nations General Assembly, December 10, 1948, https://www.un.org/en/about-us/universal-declaration-of-human-rights.

13. Ruth Macklin, "Dignity Is a Useless Concept," *BMJ* 327, no. 7429 (December 20, 2003): 1419–20, https://doi.org/10.1136%2Fbmj.327.7429.1419.

14. Steven Pinker, "The Stupidity of Dignity," *The New Republic*, May 28, 2008, https://newrepublic.com/article/64674/the-stupidity-dignity.

15. Tom L. Beauchamp and James F. Childress, *Principles of Biomedical Ethics*, 8th ed. (Oxford University Press, 2019), 67.

16. K. Danner Clouser, "A Critique of Principlism," *The Journal of Medicine & Moral Philosophy* 15, no. 2 (April 1, 1990): 219–36, https://doi.org/10.1093/jmp/15.2.219.

17. John Evans, "A Critical History of Bioethics," in *The Disability Bioethics Reader*, eds. Joel Michael Reynolds and Christine Wieseler (Routledge, 2022).

18. John H. Evans, "A Sociological Account of the Growth of Principlism," *The Hastings Center Report* 30, no. 5 (October 2000): 31–39, https://doi.org/10.2307/3527886.

19. John H. Evans, "The Social Context of Religion in the Jurisdictions of Bioethics," *Bioethics Today* (blog), November 18, 2020, https://bioethicstoday.org/blog/the-social-context-of-religion-in-the-jurisdictions-of-bioethics/#.

20. Evans, "The Social Context of Religion in the Jurisdictions of Bioethics."

21. American Medical Association, "Opinion 10.7: Ethics Committees in Health Care Institutions," Code of Medical Ethics, accessed July 21, 2024, https://code-medical-ethics.ama-assn.org/ethics-opinions/ethics-committees-health-care-institutions.

22. Ralf Jox and Stella Reiter-Theil, "Historical Development of Clinical Ethics Consultation in Europe," in *Thorny Issues in Clinical Ethics Consultation: North American and*

European Perspectives, eds. Katherine Wasson and Mark Kuczewski, vol. 143, Philosophy and Medicine (Springer Cham, 2022), 258, https://doi.org/10.1007/978-3-030-91916-0.

23. H. Tristam Engelhardt, *The Foundations of Bioethics*, 2nd ed. (Oxford University Press, 1996), https://global.oup.com/academic/product/the-foundations-of-bioethics-9780195057362.

24. Edmund Pellegrino and David C. Thomasma, *A Philosophical Basis of Medical Practice: Toward a Philosophy and Ethic of the Healing Professions* (Oxford University Press, 1981).

25. Bernard Gert, Charles M. Culver, and K. Danner Clouser, *Bioethics: A Return to Fundamentals*, 1st ed. (Oxford University Press, 1997).

26. Henry S. Richardson, "Specifying, Balancing, and Interpreting Bioethical Principles," *The Journal of Medicine and Philosophy: A Forum for Bioethics and Philosophy of Medicine* 25, no. 3 (January 1, 2000): 285–307, https://doi.org/10.1076/0360-5310(200006)25:3;1-H;FT285.

27. Albert R. Jonsen and Stephen Toulmin, *The Abuse of Casuistry: A History of Moral Reasoning*, 1st ed. (University of California Press, 1990).

28. Susan Wolf, ed., *Feminism & Bioethics: Beyond Reproduction*, 1st ed. (Oxford University Press, 1996).

29. Helga Kuhse, *Caring: Nurses, Women and Ethics* (Wiley-Blackwell, 1997).

30. Tom L. Beauchamp, "The Origins and Drafting of the Belmont Report," *Perspectives in Biology and Medicine* 63, no. 2 (Spring 2020): 245, https://doi.org/10.1353/pbm.2020.0016.

31. M. Therese Lysaught, "Respect: Or, How Respect for Persons Became Respect for Autonomy," *Journal of Medicine and Philosophy* 29, no. 6 (2004): 665–80, https://doi.org/10.1080/03605310490883028.

32. Evans, "The Social Context of Religion in the Jurisdictions of Bioethics."

33. Charles E. Rosenberg, "Meanings, Policies, and Medicine: On the Bioethical Enterprise and History," *Daedalus* 128, no. 4 (Fall 1999): 27–46.

34. LeRoy Walters, "Commissions and Bioethics," *The Journal of Medicine and Philosophy: A Forum for Bioethics and Philosophy of Medicine* 14, no. 4 (August 1, 1989): 363–68, https://doi.org/10.1093/jmp/14.4.363.

35. Eric M. Meslin and Summer Johnson, "National Bioethics Commissions and Research Ethics," in *The Oxford Textbook of Clinical Research Ethics*, eds. Ezekiel J. Emanuel et al. (University Press, 2011), 848.

36. Sheila Jasanoff, *Designs on Nature: Science and Democracy in Europe and the United States* (Princeton University Press, 2007), 173.

37. Evans, "A Sociological Account of the Growth of Principlism."

38. I was a staff member for the Clinton radiation experiments committee, his bioethics advisory commission, as well as the Obama commission.

39. Ronald Bayer, "Ethics, Politics, and Access to Health Care: A Critical Analysis of the President's Commission for the Study of Ethical Problems in Medicine and Biomedical and Behavioral Research," *Cardozo Law Review* 6, no. 2 (1984): 303–20, https://larc.cardozo.yu.edu/clr/vol6/iss2/6.

40. Margaret Fletcher Stack, "A Once and Future Biomedical Ethics Board," *The Hastings Center Report* 18, no. 2 (May 1988): 2–3.

41. Jonathan D. Moreno, *Deciding Together: Bioethics and Moral Consensus*, 1st ed. (Oxford University Press, 1995).

42. Duncan Wilson, *The Making of British Bioethics*, 1st ed. (Manchester University Press, 2014), 126.

43. The Warnock Committee, "Warnock Committee Recommendations," *British Medical Journal* 289 (July 28, 1984): 238, https://www.bmj.com/content/bmj/289/6439/238.full.pdf.

44. Wilson, *The Making of British Bioethics*, 220.

45. Ruth Shalit, "When We Were Philosopher Kings: The Rise of the Medical Ethicist," *New Republic*, April 28, 1997, https://pubmed.ncbi.nlm.nih.gov/11660448/.

46. Gisela Klinkhammer, "Deutscher Ethikrat: Der Regierung genehm," *Deutsches Ärzteblatt* 103, no. 46 (2006): A-3061 / B-2665 / C-2557, https://www.aerzteblatt.de/archiv/53491/Deutscher-Ethikrat-Der-Regierung-genehm.

47. Patrik Hummel et al., "Taking Stock of the Availability and Functions of National Ethics Committees Worldwide," *BMC Medical Ethics* 22, no. 56 (May 10, 2021), https://doi.org/10.1186/s12910-021-00614-6.

48. National Bioethics Advisory Commission, *National Bioethics Advisory Commission 1998–1999 Biennial Report* (National Bioethics Advisory Commission, 2000), https://bioethicsarchive.georgetown.edu/nbac/pubs/Biennial98-99.pdf.

49. Hellenic Republic National Commission for Bioethics and Technoethics, "Terms of Reference," The National Commission for Bioethics and Technoethics, accessed July 9, 2024, https://bioethics.gr/en/page/armodiothtes-13.

50. National Center for Etik, "National Center for Etik," Our Purpose, accessed July 9, 2024, https://nationaltcenterforetik.dk/.

51. Pavel Dmitrievich Tishchenko, "Гуманитарный Анализ Биотехнологических Проектов «улучшения» Человека ('Humanitarian Analysis of Biotechnological Projects for "Improving" Humanity')" (Moscow: Federal State Budgetary Scientific Institution,

Institute of Philosophy of the Russian Academy of Sciences, 2017), https://rscf.ru/project/15-18-30057/.

52. Nataliya Shok, "From 'Bioethics' to 'Christian Bioethics': Significance of H.T. Engelhardt's Legacy in Today's Russia," *Gosudarstvo, Religii͡a, T͡Serkov v Rossii i Za Rubezhom* 38, no. 4 (January 2020): 28, http://dx.doi.org/10.22394/2073-7203-2020-38-4-7-43.

53. Barbara Culliton, "Recombinant DNA: Cambridge City Council Votes Moratorium," *Science* 193, no. 4250 (July 23, 1976): 300–301, https://doi.org/10.1126/science.11643322.

54. Paul Berg, "Asilomar 1975: DNA Modification Secured," *Nature* 455 (September 17, 2008): 290–91, https://doi.org/10.1038/455290a.

55. Donald S. Fredrickson, "Asilomar and Recombinant DNA: The End of the Beginning," in *Biomedical Politics*, ed. Kathi E. Hanna (National Academy Press, 1991), https://www.ncbi.nlm.nih.gov/books/NBK234217/.

56. Shobita Parthasarathy, "Viewing Asilomar from the Global South," *Science* (January 31, 2025), https://www.science.org/doi/10.1126/science.adw2511.

57. Jonathan E. LoTempio and Jonathan D. Moreno, "Overcoming Challenges Associated with Broad Sharing of Human Genomic," *Nature Genetics* (January 22, 2025), https://www.nature.com/ng/articles?type=perspective&year=2025.

Chapter 6

1. Michael Toomim, "Invisible College," Invisible College: A Forum for Independent Research, n.d., 11 February 2024, https://invisible.college/.

2. John Cohen, "Did CRISPR Help—or Harm—the First-Ever Gene-Edited Babies?" *Science*, August 1, 2019, https://www.science.org/content/article/did-crispr-help-or-harm-first-ever-gene-edited-babies.

3. David Shaw, "The Consent Form in the Chinese CRISPR Study: In Search of Ethical Gene Editing," *Journal of Bioethical Inquiry* 17 (January 3, 2020): 5–10, https://doi.org/10.1007%2Fs11673-019-09953-x.

4. David Cyranoski and Heidi Ledford, "Genome-Edited Baby Claim Provokes International Outcry," *Nature News*, November 26, 2018, https://www.nature.com/articles/d41586-018-07545-0.

5. Joy Zhang, Sonia Ben Ouagrham-Gormley, and Kathleen M. Vogel, "Chinese Academics Are Becoming a Force for Good Governance," *Issues in Science and Technology* 39, no. 4 (July 2023): 81–84, https://doi.org/10.58875/BCZQ1715.

6. John Cohen, "As Creator of 'CRISPR Babies' Nears Release from Prison, Where Does Embryo Editing Stand?," *Science*, March 21, 2022.

7. Jane Qiu, "Chinese Government Funding May Have Been Used for 'CRISPR Babies' Project, Documents Suggest," *STAT*, February 25, 2019, https://www.statnews.com/2019/02/25/crispr-babies-study-china-government-funding/.

8. Jon Cohen, "Inside the Circle of Trust," *Science* 365, no. 6452 (August 2, 2019): 430–37, accessed June 21, 2022, https://www.science.org/doi/10.1126/science.365.6452.430; https://science.sciencemag.org/content/sci/365/6452/430.full.pdf.

9. Henry T. Greely, "CRISPR'd Babies: Human Germline Genome Editing in the 'He Jiankui Affair,'" *Journal of Law and the Biosciences* 6, no. 1 (October 2019): 111–83, https://doi.org/10.1093/jlb/lsz010.

10. Preetika Rana and Wenxin Fan, "Chinese Gene-Editing Experiment Loses Track of Patients, Alarming Technology's Inventors," *The Wall Street Journal*, December 18, 2018.

11. Antonio Regaldo, "Chinese Scientists Have Put Human Brain Genes in Monkeys—and Yes, They May Be Smarter," *MIT Technology Review*, April 10, 2019, https://www.technologyreview.com/2019/04/10/136131/chinese-scientists-have-put-human-brain-genes-in-monkeysand-yes-they-may-be-smarter/.

12. "Early Results from First-In-U.S. Trial of CRISPR-Edited Immune Cells for Cancer Patients Suggest Safety of Approach," *Penn Medicine News*, November 6, 2019.

13. For a thorough analysis of the case and the ethical issues more generally see Henry T. Greely, *CRISPR People: The Science and Ethics of Editing Humans* (MIT Press, 2021).

14. Yuting Zhu and Jonathan D. Moreno, "From Gene-Edited Embryos to Covid: China Faces Regulatory and Ethical Challenges," *The Hasting Center* (blog), July 31, 2022, https://www.thehastingscenter.org/from-gene-edited-embryos-to-covid-china-faces-regulatory-and-ethical-challenges/.

15. Mei Mei Chu and Laurie Chen, "China Passes Law Granting Communist Party More Control over Cabinet," *Reuters*, March 11, 2024, https://www.reuters.com/world/china/china-passes-law-granting-communist-party-more-control-over-cabinet-2024-03-11/.

16. Dennis Normile, "Chinese Scientist Who Produced Genetically Altered Babies Sentenced to 3 Years in Jail | Science | AAAS," accessed December 16, 2024, https://perma.cc/9WSN-XADZ.

17. Antonio Regalado, "The Creator of the CRISPR Babies Has Been Released from a Chinese Prison," April 4, 2022, https://www.technologyreview.com/2022/04/04/1048829/he-jiankui-prison-free-crispr-babies/.

18. Megan Molteni, "After Three Years in Prison, 'CRISPR Babies' Scientist Is Attempting a Comeback," November 29, 2022, https://www.statnews.com/2022/11/29/after-prison-crispr-babies-scientist-is-attempting-comeback/.

19. Rick Berke, "Alarmed by New 'CRISPR Babies' Plan, Top Science Figures Say They're Powerless to Stop It," June, 19, 2019 edition, accessed June 21, 2022, https://www.statnews.com/2019/06/24/outraged-by-new-crispr-babies-plan-top-science-figures-say-theyre-powerless-to-stop-it/.

20. *Nature News*. Accessed April 10, 2025. https://www.nature.com/nature-index/country-outputs/China.

21. President's Commission for the Study of Ethical Problems and in Medicine and Biomedical and Behavioral Research, *Splicing Life: A Report on the Social and Ethical Issues of Genetic Engineering with Human Beings* (National Information Resource on Ethics and Human Genetics, November 1982), http://hdl.handle.net/10822/548744.

22. In the years since *Splicing Life*, international governmental organizations such as UNESCO have published guidance documents on the human genome and human rights (1997) and human genetic data (2003), reflecting similar topics, principles, and concerns.

23. Committee on Human Gene Editing: Scientific, Medical, and Ethical Considerations, and National Academies of Sciences, Engineering, and Medicine, *Human Genome Editing: Science, Ethics, and Governance.* (The National Academies Press, 2017), https://doi.org/10.17226/24623; https://nap.nationalacademies.org/read/24623/chapter/1#xi.

24. Nuffield Council on Bioethics, *Genome Editing and Human Reproduction: Social and Ethical Issues* (Nuffield Council on Bioethics, July 2018), https://www.nuffieldbioethics.org/assets/pdfs/Genome-editing-and-human-reproduction-short-guide.pdf.

25. Cohen, "Inside the Circle of Trust," 432.

26. Li-hua Chung, "Action Urged After FBI Arrests Scientists," *Taipei Times*, October 14, 2018, https://www.taipeitimes.com/News/front/archives/2018/10/14/2003702336.

27. Mara Hvistendahl, "Exclusive: Major U.S. Cancer Center Ousts 'Asian' Researchers After NIH Flags Their Foreign Ties," *ScienceInsider*, April 19, 2019, https://www.science.org/content/article/exclusive-major-us-cancer-center-ousts-asian-researchers-after-nih-flags-their-foreign.

28. Andrew Joseph, "NIH Has Referred 16 Allegations of Foreign Influence on U.S. Research to Investigators," *STAT*, June 5, 2019, https://www.statnews.com/2019/06/05/nih-has-referred-16-allegations-of-foreign-influence-on-u-s-research-to-investigators/.

29. David Cyranoski, "Russian Biologist Plans More CRISPR-Edited Babies," *Nature* 570 (2019): 145–46, https://doi.org/10.1038/d41586-019-01770-x.

30. Jon Cohen, "Embattled Russian Scientist Sharpens Plans to Create Gene-Edited Babies," *Science*, October 21, 2019, https://www.science.org/content/article/embattled-russian-scientist-sharpens-plans-create-gene-edited-babies.

31. Jeffrey Ding and Paul Triolo, "Translation: Excerpts from China's 'White Paper on Artificial Intelligence Standardization,'" *New America: Cybersecurity Initiative* (blog), June 20, 2018, https://www.newamerica.org/cybersecurity-initiative/digichina/blog/translation-excerpts-chinas-white-paper-artificial-intelligence-standardization/.

32. Jennifer Huergo, *NIST Releases Draft Plan for Federal Engagement in AI Standards Development* (The U.S. Department of Commerce's National Institute of Standards and Technology, July 2, 2019), https://www.nist.gov/news-events/news/2019/07/nist-releases-draft-plan-federal-engagement-ai-standards-development.

33. Robert MacMillan, Dustin Volz, and Aruna Viswanatha, "China Is Stealing AI Secrets to Turbocharge Spying, U.S. Says," *The Wall Street Journal*, December 25, 2023, https://www.wsj.com/tech/ai/china-is-stealing-ai-secrets-to-turbocharge-spying-u-s-says-00413594.

34. Dyani Lewis, "Unethical Studies on Chinese Minority Groups Are Being Retracted—But Not Fast Enough, Critics Say," *Nature* 625 (2024): 650–54, https://doi.org/10.1038/d41586-024-00170-0.

35. David B. Resnik, Adil Shamoo, and Sheldon Krimsky, "Fraudulent Human Embryonic Stem Cell Research in South Korea: Lessons Learned," *Accountability in Research* 13, no. 1 (2006): 101–9, https://doi.org/10.1080%2F08989620600634193.

36. Jonathan H. Marks, "Beyond Disclosure: Developing Law and Policy to Tackle Corporate Influence," *American Journal of Law & Medicine* 46, no. 2–3 (July 12, 2020): 275–96, https://journals.sagepub.com/doi/abs/10.1177/0098858820933499?journalCode=jlma.

37. Jonathan D. Moreno, *Undue Risk: Secret State Experiments on Humans* (W.H. Freeman, 1999).

38. Whitney Kerr and Harald Schmidt, "COVID-19 Ventilator Rationing Protocols: Why We Need to Know More About the Views of Those with Most to Lose," *Journal of Medical Ethics*, K, 47 (2021): 133–36, https://doi.org/10.1136/medethics-2020-106948.

39. Harald Schmidt et al., "Equitable Allocation of COVID-19 Vaccines in the United States," *Nature Medicine* 27 (July 2021): 1298–1307, https://doi.org/10.1038/s41591-021-01379-6.

40. Jonathan H. Marks, "Trust in Crises and Crises of Trust," *The Hastings Center Report* 53, no. 22 (October 2023): S9–15, https://doi.org/10.1002/hast.1518.

Chapter 7

1. Henk A.M.J. ten Have and Bert Gordijn, "Global Bioethics," in *Handbook of Global Bioethics*, eds. Henk A.M.J. ten Have and Bert Gordijn, vol. 1 (Springer, 2013), 16.

2. ten Have and Gordijn, "Global Bioethics."

3. ten Have and Gordijn, p. 14.

4. Jonathan D. Moreno, *The Body Politic: The Battle Over Science in America* (Bellevue Literary Press, 2011).

5. "Guidelines for Stem Cell Research and Clinical Translation," International Society for Stem Cell Research, 2021, https://www.isscr.org/guidelines.

6. D. H. Guston, "Preface," in *The Yearbook of Nanotechnology in Society: Presenting Futures*, eds. Erik Fisher, Cynthia Selin, and Jameson M. Wetmore, vol. 1 (Springer, 2008).

7. Dorit Barlevy et al., "Governing with Public Engagement: An Anticipatory Approach to Human Genome Editing," *Science and Public Policy*, March 25, 2024, https://doi.org/10.1093/scipol/scae010.

8. George W. Bush, "US National Security Strategy: Overview of America's International Strategy," June 1, 2002, https://2001-2009.state.gov/r/pa/ei/wh/15421.htm

9. Joe Biden, "National Security Strategy," October 12, 2022, https://www.whitehouse.gov/wp-content/uploads/2022/10/Biden-Harris-Administrations-National-Security-Strategy-10.2022.pdf.

10. Edward P. Richards, Katharine C. Rathbun and Jay Alexander Gold, "The Smallpox Vaccination Campaign of 2003: Why Did it Fail and What are the Lessons for Bioeterrorism Preparedness?" *Louisiana Law Review* 64: 851–904, 2004. https://www.researchgate.net/publication/228125599_The_Smallpox_Vaccination_Campaign_of_2003_Why_Did_it_Fail_and_What_are_the_Lessons_for_Bioterrorism_Preparedness.

11. David Malakoff, "Rough Entry for Foreign Scholars," *Science*, December 16, 2002, https://www.science.org/content/article/rough-entry-foreign-scholars

12. Yojana Sharma, "US Targets Chinese Talent in Drive to 'Decouple' Science," *University World News*, December 12, 2020, https://www.universityworldnews.com/post.php?story=20201211141413735.

13. Center for Strategic and International Studies, *Survey of Chinese Espionage in the United States Since 2000* (Center for Strategic and International Studies, March 2023), https://www.csis.org/programs/strategic-technologies-program/survey-chinese-espionage-united-states-2000.

14. Jay Patel and Devi Sridhar, "Toward Better Pandemic Preparedness," *Finance and Development*, December 2021, https://www.imf.org/en/Publications/fandd/issues/2021/12/Pandemic-preparedness-Patel-Sridhar.

15. Sheryl Gay Stohlberg, Chris Hamby, and Sharon LaFraniere, "Emergent Hid Evidence of Covid Vaccine Problems at Plant, Report Says," *New York Times*, May 10,

2022, https://www.nytimes.com/2022/05/10/us/politics/emergent-fda-vaccine-covid-contaminated.html.

16. John A. Muckstadt et al., "Efficient and Effective Large-Scale Vaccine Distribution," *International Journal of Production Economics* 262 (August 2023), https://www.sciencedirect.com/science/article/pii/S0925527323001536.

17. United Nations Economic and Social Council, "Unequal Vaccine Distribution Self-Defeating, World Health Organization Chief Tells Economic and Social Council's Special Ministerial Meeting," 2021 Session, Special Meeting on "A Vaccine for All," April 16, 2021, https://press.un.org/en/2021/ecosoc7039.doc.htm.

18. Elcano Royal Institute, "COVID-19: The Geopolitics of the Vaccine, a Weapon for Global Security," March 15, 2021, https://www.realinstitutoelcano.org/en/analyses/covid-19-the-geopolitics-of-the-vaccine-a-weapon-for-global-security/.

19. Jon Henley, "Is Russia's Covid Vaccine Anything More than a Political Weapon?" *The Guardian*, April 30, 2021, https://www.theguardian.com/society/2021/apr/30/is-russias-covid-vaccine-anything-more-than-a-political-weapon-sputnik-v.

20. Jonathan D. Moreno, Judit Sándor, and Ulf Schmidt, "The Vaccination Cold War," *The Hastings Center Report* 51, no. 5 (September 16, 2021): 12–17, https://doi.org/10.1002/hast.1282.

21. Sergio G. Litewka and Jonathan D. Moreno, "Covid-19 in Argentina and the Abuse of Bioethics," May 20, 2021, https://www.thehastingscenter.org/covid-19-in-argentina-and-the-abuse-of-bioethics/.

22. Chris Bing and Joel Schectman, "Pentagon Ran Secret Anti-Vax Campaign to Undermine China during the Pandemic," *Reuters*, June 14, 2024, https://www.reuters.com/investigates/special-report/usa-covid-propaganda/.

23. Jonathan E. LoTempio et al., "We Can Do Better: Lessons Learned on Data Sharing in COVID-19 Pandemic Can Inform Future Outbreak Preparedness and Response," Science & Diplomacy, August 19, 2020, https://www.sciencediplomacy.org/article/2020/we-can-do-better-lessons-learned-data-sharing-in-covid-19-pandemic-can-inform-future.

24. Joe McDonald and Huizhong Wu, "Top Chinese Official Admits Sinovac Effectiveness Is Low," *AP News*, April 11, 2021, https://apnews.com/article/china-gao-fu-vaccines-offer-low-protection-coronavirus-675bcb6b5710c7329823148ffbff6ef9.

25. Penny Bailey, "Hungary's Cold War Battle with Polio," *BBC*, April 14, 2014, https://www.bbc.com/future/article/20140415-how-polio-eased-cold-war-tensions.

26. Moreno, Sándor, and Schmidt, "The Vaccination Cold War," 13.

27. Ezekiel J. Emanuel et al., "On the Ethics of Vaccine Nationalism: Framework," *Ethics & International Affairs* 35, no. 4 (Winter 2021): 543–62, https://pmc.ncbi.nlm.nih.gov/articles/PMC7893355/ doi: 10.2105/AJPH.2020.306098

28. Paul Stronski, "What Went Wrong With Russia's Sputnik V Vaccine Roll-out?," Carnegie Endowment for International Peace, November 15, 2021, https://carnegieendowment.org/posts/2021/11/what-went-wrong-with-russias-sputnik-v-vaccine-rollout?lang=en.

29. Nataliya Shok, "Health and Security in Russia: Putin's First Special Operation," Kenan Cable, July 2024, 93 editions, https://www.wilsoncenter.org/publication/kennan-cable-no-93-health-and-security-russia-putins-first-special-operation.

30. Reynald Castañeda, "Clinical Trials in Russia: Big Pharma Makes Moves but What's the Pipeline Impact?," *Clinical Trials Arena*, March 24, 2022, https://www.clinicaltrialsarena.com/ukraine-crisis/clinical-trials-in-russia-big-pharma-makes-moves/.

31. Volker Hess, Laura Hottenrott, and Peter Steinkamp, *Testen im Osten. DDR-Arzneimittelstudien im Auftrag westlicher Pharmaindustrie, 1964–1990, [Testing in the East. GDR Drug Studies on Behalf of the Western Pharmaceutical Industry]* (be.bra wissenschaft, 2016).

32. Dóra Vargha, "Vaccination and the Communist State: Polio in Eastern Europe," in *The Politics of Vaccination: A Global History*, eds. Christine Holmberg, Stuart Blume, and Paul Greenough, 1st ed., vol. 9, Social History of Medicine (Manchester University Press, 2017), 360.

33. World Health Organization, "WHO Model List of Essential Medicines," 23rd List, 2023, July 26, 2023, https://www.who.int/publications/i/item/WHO-MHP-HPS-EML-2023.02.

34. United Nations Security Council, "Adopting Resolution 2664 (2022), Security Council Approves Humanitarian Exemption to Asset Freeze Measures Imposed by United Nations Sanctions Regimes," 9214th Meeting (AM), December 9, 2022, https://press.un.org/en/2022/sc15134.doc.htm.

35. Isabel van Brugen, "Russia's Medicine Shortage Worsens as Sanctions Take Toll," *Newsweek*, September 19, 2023, https://www.newsweek.com/russia-medicine-shortage-sanctions-ukraine-war-1828118.

36. Nicole Wetsman, "Russian Government Bars its Scientists from International Conferences," *The Verge*, March 21, 2022, https://www.theverge.com/2022/3/21/22988994/russia-science-publication-database-conferences.

37. Ministry of Health of the Russian Federation, Order 200n, "On Approval of Rules for Good Clinical Practice," April 1, 2016, registered in the Ministry of Justice, August 23, 2016.

38. Petr Talantov, Ravil Niyazov, Galina Viryasova, Margarita Dranitsyna and Ilya Yasqny, "Unapproved Clinical Trials in Russia: Exception or Norm?," *BMC Medical Ethics* 22, no. 1 (2021): doi:10.1186/s12910-021-00617-3.

39. Donalo Paolo Mancini and Henry Foy, "EU Regulator to Probe Ethical Standards of Sputnik Vaccine Trials," *Financial Times*, April 7, 2021, https://www.ft.com/content/50031165-1f46-446b-be9a-36d553805fec.

40. Denia Y. Logunov, Inna V. Dolzhikova, Olga V. Zubkova, Amir I. Tukhvatulin, Dmitry Shcheblyyakov, Alina Dzharullaeva et al., "Safety and Immunogenicity of an rAd26 and rAd5 Vector-Based Heterologous Prime-Boost COVID-19 Vaccine in Two Formulations: Two Open, NonrandomizedPhase 1/2 Studies from Russia," *Lancet* 396 (2020): doi:10.1016/S0140-6736(20)31866-3.

41. Denia Y. Logunov, Inna V. Dolzhikova, Olga V. Zubkova, Amir I. Tukhvatulin, Dmitry Shcheblyyakov, Alina Dzharullaeva et al., "Safety and Efficacy of an rAd26 and rAd5 Vector-Based Heterologous Prime-Boost COVID-19 Vaccine: An Interim Analysis of a Randomised Controlled Phase 3 Trial in Russia," *Lancet* 397 (2021): doi:10.1016/S0140-6736(21)00234-8.

42. Enrico M. Bucci, Johannes Berkhof, Andre Gillibert, Gowri Gopalakrishna, Raffaele A. Calogero, Lex M. Bouter et al., "Data Discrepancies and Substandard Reporting of Interim Dataof Sputnik V Phase 3 Trial," letter to the editor, *Lancet* 397 (2021): 1881-83, https://www.thelancet.com/journals/lancet/article/PIIS0140-6736(21)00899-0/fulltext.

43. Golfo Alexopoulos, "Medical Research in Stalin's Gulag," *Bulletin of the History of Medicine* 90, no. 3 (Fall 2016): 363–93, https://doi.org/10.1353/bhm.2016.0070.

44. Vasiliy Vlassov, "Russian Medicine and the Nuremberg Trials," *European Journal of Public Health* 16, no. 3 (June 2006): 229, https://doi.org/10.1093/eurpub/ckl038.

45. Commission of the Russian Academy of Sciences on Combating Pseudoscience and Falsification of Scientific Research, *Position of Russian Academy of Science Specialists on the Collection, Storage, Use, and Transfer of Human Biomaterials* (Commission of the Russian Academy of Sciences on Combating Pseudoscience and Falsification of Scientific Research, November 13, 2017), https://www.ras.ru/FStorage/Download.aspx?id=ac12f872-e0bb-46a5-a68b-cd107610b4b4.

46. Richard Stone, "'A Step to Nowhere': Russian Scientist Organizes Protest of Ukraine War," *Science*, February 24, 2022, https://www.science.org/content/article/step-nowhere-russian-scientist-organizes-protest-ukraine-war.

47. Karolina Zhukoff, "The Danger of Conflating the Russian Scientific Community with Putin's Regime in Global Bioethics," *Bioethics Today*, December 13, 2024, https://bioethicstoday.org/blog/the-danger-of-conflating-the-russian-scientific-community-with-putins-regime-in-global-bioethics/.

48. Joseph J. Fins, "Bioethics, Ukraine, and the Peril of Silence," *Cambridge Quarterly of Healthcare Ethics* 32, no. 1 (July 8, 2022): 1–3, https://doi.org/10.1017/S0963180122000238.

49. György Péteri, "Nylon Curtain—Transnational and Transsystemic Tendencies in the Cultural Life of State-Socialist Russia and East-Central Europe," *Slavonica* 10, no. 2 (2004): 113–23, https://doi.org/10.1179/sla.2004.10.2.113.

50. Judit Sándor, "Bioethics as a Tool for Extending the Human Rights Discourse," in *Human Rights in the 21st Century*, eds. Tibor Várady and Miodrag Jovanovic (The Hague, Netherlands: Eleven International Publishing, 2020), 348.

51. World Health Organization, "Working Group on Developing WHO Guidance on Clinical Ethics," https://www.who.int/groups/working-group-on-developing-who-guidance-on-clinical-ethics.

52. Jonathan H. Marks, "Toward a Unified Theory of Professional Ethics and Human Rights," *Michigan Journal of International Law* 33, no. 2 (n.d.), https://core.ac.uk/download/pdf/232683861.pdf.

53. Madison Powers and Ruth Faden, *Social Justice: The Moral Foundations of Public Health and Health Policy*, 1st Edition, Issues in Biomedical Ethics (Oxford University Press, 2006).

54. Stephanie Nixon and Lisa Forman, "Exploring Synergies Between Human Rights and Public Health Ethics: A Whole Greater than the Sum of Its Parts," *BMC International Health and Human Rights* 8, no. 2 (2008), https://doi.org/10.1186/1472-698X-8-2.

55. "Siracusa Principles: On the Limitation and Derogation Provisions in the International Covenant on Civil and Political Rights," American Association for the International Commission of Jurists, April 1985, https://www.icj.org/wp-content/uploads/1984/07/Siracusa-principles-ICCPR-legal-submission-1985-eng.pdf.

56. Nicholas G. Evans, *War on All Fronts: A Theory of Health Security Justice*, 1st ed. (MIT Press, 2023), https://mitpress.mit.edu/9780262545433/war-on-all-fronts/.

57. James F. Childress et al., "Public Health Ethics: Mapping the Terrain," *Journal of Law and Medical Ethics* 30, no. 2 (2002): 170–78, https://doi.org/10.1111/j.1748-720x.2002.tb00384.x.

58. R. E. G. Upshur, "Principles for the Justification of Public Health Intervention," *Canadian Journal of Public Health* 93, no. 2 (March 1, 2002): 101–3, https://doi.org/10.1007/bf03404547.

59. Robert Baker, "Bioethics and Human Rights: A Historical Perspective," *Cambridge Quarterly of Healthcare Ethics* 10, no. 3 (June 29, 2001): 213, https://doi.org/10.1017/S0963180101003048.

60. Gov.UK, Guidance: Pandemic Flu, 24 November 2017, https://www.gov.uk/guidance/pandemic-flu#ethical-framework.

61. Baker, "Bioethics and Human Rights," 249.

62. Angus Dawson, "Resetting the Parameters: Public Health as the Foundation for Public Health Ethics," in *Public Health Ethics: Key Concepts and Issues in Policy and Practice*, ed. Angus Dawson (Cambridge University Press, 2011), 12, https://doi.org/10.1017/CBO9780511862670.

63. Dawson, "Resetting the Parameters."

64. Donella H. Meadows et al., *The Limits to Growth: A Report for the Club of Rome's Project on the Predicament of Mankind* (Universe Pub, 1972), https://www.library.dartmouth.edu/digital/digital-collections/limits-growth.

65. Peter J. Whitehouse, In Memoriam. Van Rensselaer Potter: The Original Bioethicist," *Global Bioethics*, 14(4), 2001: 47-48, https://www.tandfonline.com/doi/abs/10.1080/11287462.2001.10800814.

66. EU General Data Protection Regulation (GDPR): Regulation (EU) 2016/679 of the European Parliament and of the Council of 27 April 2016 on the protection of natural persons with regard to the processing of personal data and on the free movement of such data, and repealing Directive 95/46/EC (General Data Protection Regulation), OJ L 119, 04.05.2016; cor. OJ L 127, 23.5.2018. Art 4(1).

67. Ciara Staunton, Santa Slokenberga, and Deborah Mascalzoni, "The GDPR and the Research Exemption: Considerations on the Necessary Safeguards for Research Biobanks," *European Journal of Human Genetics* 27 (April 17, 2019): 1159–67, https://doi.org/10.1038/s41431-019-0386-5.

68. Paul M. Schwartz, "Global Data Privacy: The EU Way," *NYU Law Review* 94, no. 4 (October 10, 2019), https://www.nyulawreview.org/wp-content/uploads/2019/10/NYULAWREVIEW-94-4-Schwartz.pdf.

69. Ruth R. Faden et al., "An Ethics Framework for a Learning Health Care System: A Departure from Traditional Research Ethics and Clinical Ethics," *The Hastings Center Report* 43, no. s1 (January 11, 2013): S16–27, https://doi.org/10.1002/hast.134.

70. Emily A. Largent, Steven Joffe, and Franklin G. Miller, "Can Research and Care Be Ethically Integrated?," *The Hastings Center Report* 21, no. 4 (February 8, 2012): 37–46, https://doi.org/10.1002/j.1552-146X.2011.tb00123.x.

71. Staunton, Slokenberga, and Mascalzoni, "The GDPR and the Research Exemption."

72. "China's Enormous Surveillance State Is Still Growing: Its Citizens Don't Seem to Mind," *The Economist*, November 23, 2023, https://www.economist.com/china/2023/11/23/chinas-enormous-surveillance-state-is-still-growing.

Index

Basic Bioethics

Arthur Caplan, editor

Books Acquired under the Editorship of Glenn McGee and Arthur Caplan

Peter A. Ubel, *Pricing Life: Why It's Time for Health Care Rationing*

Mark G. Kuczewski and Ronald Polansky, eds., *Bioethics: Ancient Themes in Contemporary Issues*

Suzanne Holland, Karen Lebacqz, and Laurie Zoloth, eds., *The Human Embryonic Stem Cell Debate: Science, Ethics, and Public Policy*

Gita Sen, Asha George, and Piroska Östlin, eds., *Engendering International Health: The Challenge of Equity*

Carolyn McLeod, *Self-Trust and Reproductive Autonomy*

Lenny Moss, *What Genes Can't Do*

Jonathan D. Moreno, ed., *In the Wake of Terror: Medicine and Morality in a Time of Crisis*

Glenn McGee, ed., *Pragmatic Bioethics, 2d edition*

Timothy F. Murphy, *Case Studies in Biomedical Research Ethics*

Mark A. Rothstein, ed., *Genetics and Life Insurance: Medical Underwriting and Social Policy*

Kenneth A. Richman, *Ethics and the Metaphysics of Medicine: Reflections on Health and Beneficence*

David Lazer, ed., *DNA and the Criminal Justice System: The Technology of Justice*

Harold W. Baillie and Timothy K. Casey, eds., *Is Human Nature Obsolete? Genetics, Bioengineering, and the Future of the Human Condition*

Robert H. Blank and Janna C. Merrick, eds., *End-of-Life Decision Making: A Cross-National Study*

Norman L. Cantor, *Making Medical Decisions for the Profoundly Mentally Disabled*

Margrit Shildrick and Roxanne Mykitiuk, eds., *Ethics of the Body: Post-Conventional Challenges*

Alfred I. Tauber, *Patient Autonomy and the Ethics of Responsibility*

David H. Brendel, *Healing Psychiatry: Bridging the Science/Humanism Divide*

Jonathan Baron, *Against Bioethics*

Michael L. Gross, *Bioethics and Armed Conflict: Moral Dilemmas of Medicine and War*

Karen F. Greif and Jon F. Merz, *Current Controversies in the Biological Sciences: Case Studies of Policy Challenges from New Technologies*

Deborah Blizzard, *Looking Within: A Sociocultural Examination of Fetoscopy*

Ronald Cole-Turner, ed., *Design and Destiny: Jewish and Christian Perspectives on Human Germline Modification*

Holly Fernandez Lynch, *Conflicts of Conscience in Health Care: An Institutional Compromise*

Mark A. Bedau and Emily C. Parke, eds., *The Ethics of Protocells: Moral and Social Implications of Creating Life in the Laboratory*

Jonathan D. Moreno and Sam Berger, eds., *Progress in Bioethics: Science, Policy, and Politics*

Eric Racine, *Pragmatic Neuroethics: Improving Understanding and Treatment of the Mind-Brain*

Martha J. Farah, ed., *Neuroethics: An Introduction with Readings*

Jeremy R. Garrett, ed., *The Ethics of Animal Research: Exploring the Controversy*

Books Acquired under the Editorship of Arthur Caplan

Sheila Jasanoff, ed., *Reframing Rights: Bioconstitutionalism in the Genetic Age*

Christine Overall, *Why Have Children? The Ethical Debate*

Yechiel Michael Barilan, *Human Dignity, Human Rights, and Responsibility: The New Language of Global Bioethics and Bio-Law*

Tom Koch, *Thieves of Virtue: When Bioethics Stole Medicine*

Timothy F. Murphy, *Ethics, Sexual Orientation, and Choices about Children*

Daniel Callahan, *In Search of the Good: A Life in Bioethics*

Robert Blank, *Intervention in the Brain: Politics, Policy, and Ethics*

Gregory E. Kaebnick and Thomas H. Murray, eds., *Synthetic Biology and Morality: Artificial Life and the Bounds of Nature*

Dominic A. Sisti, Arthur L. Caplan, and Hila Rimon-Greenspan, eds., *Applied Ethics in Mental Healthcare: An Interdisciplinary Reader*

Barbara K. Redman, *Research Misconduct Policy in Biomedicine: Beyond the Bad-Apple Approach*

Russell Blackford, *Humanity Enhanced: Genetic Choice and the Challenge for Liberal Democracies*

Nicholas Agar, *Truly Human Enhancement: A Philosophical Defense of Limits*

Bruno Perreau, *The Politics of Adoption: Gender and the Making of French Citizenship*

Carl Schneider, *The Censor's Hand: The Misregulation of Human-Subject Research*

Lydia S. Dugdale, ed., *Dying in the Twenty-First Century: Towards a New Ethical Framework for the Art of Dying Well*

John D. Lantos and Diane S. Lauderdale, *Preterm Babies, Fetal Patients, and Childbearing Choices*

Harris Wiseman, *The Myth of the Moral Brain*

Arthur L. Caplan and Jason Schwartz, eds., *Vaccine Ethics and Policy: An Introduction with Readings*

Tom Koch, *Ethics in Everyday Places: Mapping Moral Stress, Distress, and Injury*

Nicole Piemonte, *Afflicted: How Vulnerability Can Heal Medical Education and Practice*

Abigail Gosselin, *Mental Patient: Ethics from a Patient's Perspective*

Laurie Zoloth, *May We Make the World?*

Robert Baker, *Making Modern Medical Ethics*

Margaret Battin, *Sex and the Planet: What "Opt In" Reproduction Could Do for the Globe*

John R. Shook and James Giordano, *Bioethics and Brains: A Disciplined and Principled Neuroethics*

Erice Racine, *The Theory of Deliberative Wisdom*

Jonathan D. Moreno, *Absolutely Essential: Bioethics and the Rules-Based International Order*

Publisher contact:
The MIT Press
Massachusetts Institute of Technology
77 Massachusetts Avenue, Cambridge, MA 02139
mitpress.mit.edu

EU Authorised Representative:
Easy Access System Europe, Mustamäe tee 50,
10621 Tallinn, Estonia
gpsr.requests@easproject.com

Printed by Integrated Books International,
United States of America